Preface

The purpose of this *Compendium of Bean Diseases, Second Edition* is to provide an updated, comprehensive, authoritative, and modern account of bean diseases. It is international in scope and practical in emphasis. It is designed to assist in the diagnosis of bean diseases, whether in the field, laboratory, or diagnostic clinic, and to provide recommendations for management of bean diseases. The compendium should be useful to plant pathologists, crop production specialists, growers, diagnostic clinicians, students, regulatory agents, crop consultants, agribusiness representatives, educators, researchers, and others interested in the recognition or management of bean diseases throughout the world.

The compendium describes infectious diseases caused by fungi, bacteria, nematodes, viruses, and phytoplasmas and noninfectious diseases caused by abiotic factors, such as moisture stress, temperature stress, pesticides, air pollution, and mineral deficiencies and toxicities. It does not deal with insects except to the extent that they are involved in the disease discussed, for example, as vectors of viruses.

The compendium contains many illustrations of bean diseases and their causal agents. In addition, symptoms of the diseases and identifying characteristics of the causal agents are described.

The management recommendations are general in nature so that they may be adapted to a wide range of cropping conditions. Specific recommendations on chemicals or cultivars are not given because they are soon outdated, may not be generally available, or may not be applicable to certain regions. Disease management practices should be economical and must be compatible with the environment and the production system. Thus, many options for bean disease management are discussed.

The description of each disease includes a general account of its importance and world distribution, symptoms, causal organism or agent, disease cycle and epidemiology, management, and selected references. The references document the descriptions and may be consulted for further information.

This compendium resulted from the efforts of many people as authors, photographers, reviewers, and sponsors to whom we express our deepest thanks. The editors gratefully acknowledge the support, time, and facilities provided by our home institutions to this effort on behalf of the international community.

We also wish to thank the following individuals who supplied figures and photographs for this compendium.

G. S. Abawi, Cornell University, Geneva, NY

L. Bos, Research Institute for Plant Protection, Wageningen, Netherlands

M. H. Dickson, Cornell University, Geneva, NY

E. Drijfhout, Institute for Horticulural Plant Breeding, Wageningen, Netherlands

R. L. Forster, University of Idaho, Kimberly, ID

G. D. Franc, University of Wyoming, Laramie, WY

G. Godoy-Lutz, University of Nebraska, Lincoln, NE

D. J. Hagedorn, University of Wisconsin, Madison, WI

R. Hall, University of Guelph, Guelph, Ontario

R. O. Hampton, USDA-ARS, Corvallis, OR

L. E. Hanson, USDA-ARS, Fort Collins, CO

R. M. Harveson, University of Nebraska, Scotts Bluff, NE

W. R. Jarvis, Agriculture Canada, Harrow, Ontario

C. W. Kuhn, University of Georgia, Athens, GA

A. J. Liebenberg, ARC-Grain Crops Institute, Potchefstroom, Republic of South Africa

M. M. Liebenberg, ARC-Grain Crops Institute, Potchefstroom, Republic of South Africa

H. H. Lyon, Cornell University, Ithaca, NY

W. F. Mai, Cornell University, Ithaca, NY

M. S. McMillan, Colorado State University, Fort Collins, CO

S. K. Mohan, University of Idaho, Parma, ID

F. J. Morales, Centro Internacional de Agricultura Tropical, Cali, Colombia

M. Pastor-Corrales, USDA-ARS, Beltsville, MD

W. F. Pfender, USDA-ARS, Corvallis, OR

R. Provvidenti, Cornell University, Geneva, NY

A. W. Saettler, USDA-ARS, East Lansing, MI

H. F. Schwartz, Colorado State University, Fort Collins, CO

H. A. Scott, University of Arkansas, Fayetteville, AR

M. J. Silbernagel, USDA-ARS, Prosser, WA

J. R. Stavely, USDA-ARS, Beltsville, MD

J. R. Steadman, University of Nebraska, Lincoln, NE

D. R. Sumner, University of Georgia, Tifton, GA

J. C. Tu, Agriculture Canada, Harrow, Ontario

D. M. Webster, Seminis Inc., Twin Falls, ID

Howard F. Schwartz
James R. Steadman
Robert Hall
Robert L. Forster

Contributors

George Abawi
Cornell University
Geneva

Greg Boland
University of Guelph
Guelph, Ontario

Mark Brick
Colorado State University
Fort Collins

Robert Forster (retired)
University of Idaho R & E Center
Kimberly

Gary Franc
University of Wyoming
Laramie

Graciela Godoy-Lutz
University of Nebraska
Lincoln

Robert Hall
University of Guelph
Guelph, Ontario

Linda Hanson
USDA-ARS Sugarbeet Lab
Fort Collins

Robert Harveson
University of Nebraska
Scottsbluff

Carol Ishimaru
University of Minnesota
St. Paul

George Mahuku
Centro Internacional de Agricultura Tropical
Cali, Colombia

Robert McMillan
Univ. of Florida, IFAS
Homestead

Krishna Mohan
University of Idaho
Parma

Francisco Morales
Centro Internacional de Agricultura Tropical
Cali, Colombia

Marcial Pastor Corrales
USDA-ARS
Beltsville

Howard F. Schwartz
Colorado State University
Fort Collins

James R. Steadman
University of Nebraska
Lincoln

Jui-Chang Tu
Harrow Research Station, Agr. Canada
Harrow, Ontario

David Webster
Seminis Inc.
Twin Falls

Gary Yuen
University of Nebraska
Lincoln

Contents

Compendium of Bean Diseases

SECOND EDITION

Introduction

The Bean Plant

Common bean (*Phaseolus vulgaris* L.) was domesticated by Native Americans during pre-Colombian times. Archeological data suggest that bean was independently domesticated in different regions of the Americas, including the Andean region of South America, Argentina, and Mexico. The oldest domesticated beans found at archeological sites in each of these regions were estimated to have been cultivated between 7,000 and 9,600 years ago. Wild or putatively wild relatives of *P. vulgaris* grow currently from northern Mexico to Argentina, often in the same regions as cultivated forms. Domestication has altered the morphology and phenology of the plant, especially growth habit, seed size, seed retention, and maturity. Selection toward smaller, denser plants resulted in shorter internodes, suppressed climbing ability, fewer and thicker stems, and larger leaves. This selection strategy culminated in the compact growth habit of free-standing, determinate, and upright indeterminate bean cultivars that were more suitable for mechanized crop production. The most striking difference between wild ancestors and cultivated beans are changes in pod and seed size. During domestication, large seeds were selected for dry seed production in preference to the small seeds (preferred for garden bean production) and less dehiscent pods with lower pod fiber content. The large seed size of early domesticates indicates that gain from selection for large seed size was rapid rather than gradual. Seed colors, markings, and shapes vary widely in the species, and local landraces reflect regional preferences in seed type. For example, Venezuela and Guatemala favor black beans; Colombia and Honduras, red; Peru and Mexico, cream, tan, or black; and Brazil, black or tan striped. Landraces of climbing beans also occur as mixtures of seed types, especially in Africa where large-seeded, colored Andean beans are preferred.

Common bean is the third most important food legume crop worldwide; only soybean (*Glycine max* (L.) Merr.) and peanut (*Arachis hypogaea* L.) have more production. Cultivated beans are divided into two groups based on their edible parts. Dry edible beans are consumed as the mature dry seeds after rehydration, and snap beans (e.g., green, string, French, or Haricot beans) are consumed for their fleshy immature pods. Dry beans are further divided into distinct market classes based on seed characteristics, and snap bean classes are based on pod characteristics and plant type. Market classes of dry beans grown in North America include pinto, great northern, pink, small red, black, navy, small white, light red kidney, dark red kidney, yellow eye, Anasazi, and cranberry. Other bean species produced include lima bean, mung bean, and azuki bean. Snap bean classifications include green, wax, and Romano (e.g., Italian, flat pod). Both the dry seeds and fresh green pods of common bean are consumed throughout the world for their nutritional content.

Taxonomically, common bean belongs to the family Fabaceae (Leguminosae), which is further subdivided into subfamily, tribe, subtribe, and genus. The genus *Phaseolus* is a member of the subfamily Papilonoideae, tribe Phaseoleae, and subtribe Phaseolinae. The subtribe Phaseolinae includes many other important pulse crops, such as cowpea (*Vigna unguiculata* (L.) Walp. subsp. *unguicalata* (L.) Walp.), mung bean (*V. radiata* (L.) R. Wilczek var. *radiata* (L.) R. Wilczek), azuki bean (*V. angularis* (Willd.) Ohwi & H. Ohashi var. *angularis* (Willd.) Ohwi & H. Ohashi), moth bean (*V. aconitifolia* (Jacq.) Maréchal), and winged bean (*Psophocarpus tetragonolobus* (L.) DC.). Within the genus *Phaseolus*, the exact number of species is still unknown. A 1999 review of the genus by Debouck suggests that it contains 51 species. Species of the genus *Phaseolus* have been grouped into sections, based on plant morphology, hybridology, palynology, and molecular genetics, that reflect different lines of evolution and speciation. Debouck classified four sections, including Chiapasana, Phaseolus, Minkelersia, and Xanthotricha. The Phaseolus section included four of the cultivated *Phaseolus* species, namely, *P. vulgaris* (common bean), *P. coccineus* L. (runner bean), *P. lunatus* L. var. *lunatus* L. (lima bean), and *P. acutifolius* A. Gray var. *acutifolius* A. Gray (tepary bean). Each cultivated species was domesticated from wild ancestors that still grow in the neotropics.

Worldwide, *P. vulgaris* is the most widely grown of the four species. It is cultivated extensively in North, South, and Central America, Africa, Asia, and throughout Europe. According to the Food and Agricultural Organization (FAO), Brazil and Mexico are the largest *Phaseolus*-producing nations in the world, with an annual production of 138,700 and 75,000 metric tons (t), respectively, in 2001. The FAO statistics suggest that Asia, in particular India (213,000 t) and China (84,400 t), produces large quantities of dry beans; however, these are largely *Vigna* beans. Worldwide production of dry beans is approximately 11.6 million t harvested from 14.3 million ha. Data on the world production of snap beans are confounded by FAO statistics that combine pod production of common bean with *Vigna* species, which are consumed largely in India and China.

Morphologically, *P. vulgaris* is distinct from other legumes. The primary leaves are unifoliolate and subsequent leaves are trifoliolate. Flowers are borne on a pedicel in the axes of nodes of both primary and secondary branches. Flowers are perfect and have typical legume morphology, consisting of five petals, 10 stamens, a style, a stigma, and a superior ovary. The five petals of the corolla are differentiated into two fused petals that form a keel, two wing petals, and a standard. The keel is coiled into two to three spiral turns and contains one free and nine fused stamens with one pistil. The ovary typically contains five to eight ovules. Triangular-shaped stipules are present at the base of the corolla.

Bean pods are linear and have two valves. Pods may be parchmented with strong dehiscence, leathery with less dehiscence, or fleshy and stringless with little dehiscence. Plants with parchmented pods are used for dry bean production and are harvested when mature. Plants with fleshy pods are used for snap bean production and are harvested when the pods and seeds are still immature. Plants with leathery pods can be used for both dry and fresh bean production.

The seed consists of the embryo, two cotyledons, and a seed coat (testa). These parts constitute 9, 90, and 1% of the dry weight, respectively (Fig. 1). The embryo is essentially a miniature plant with three basic components, including the primary root or radicle, a plumule or shoot, and cotyledons. The cotyledons are simple leaves that serve as a food source for the developing plant during seed germination. The seed coat or testa is a thin structure, made up of several layers, that serves to protect the seed from mechanical damage and a potentially harmful environment during seed storage. The two scars visible on the incurved edge of the seed surface are the hilum and micropyle. The hilum is a scar that forms at the point of seed detachment from the pod after dehiscence. The micropyle is the remnant of the opening in the ovule where the pollen tube entered during fertilization. Both structures can serve as a site of water entry during the imbibition phase of seed germination.

Seed germination is the resumption of embryonic growth following a period of dormancy. Seed germination is initiated by the imbibition of water, followed by enzyme activation and synthesis, cell expansion, and subsequent rupture of the seed coat. The optimum temperature for germination is between 18 and 25°C. The minimum temperature for uniform germination is 12°C. Germination of common bean is considered epigeal, because the cotyledons are pushed above the soil surface. Following imbibition, the hypocotyl arch pushes the cotyledons through the soil and straightens, and the two primary (unifoliolate) leaves unfold (Fig. 2). After seedling emergence, the first trifoliolate leaf expands and the terminal meristem initiates formation of new leaves. Injury to the terminal bud during seedling development causes axillary buds to initiate growth and assume the function of the terminal bud. If the injury destroys both the terminal meristem and axillary buds at the unifoliolate leaf, the plant dies because no buds exist below that point. This type of injury is often associated with seed damage in the handling process or during seedling emergence because of mechanical injury or hail damage.

The bean plant undergoes four distinct developmental stages during its life cycle. The stages include I) emergence and early vegetative growth, II) branching and rapid vegetative growth, III) flower and pod formation, and IV) pod fill and maturation (Fig. 3; Table 1). The first two developmental stages occur during vegetative growth and the final two occur during reproductive growth. The time period required to complete each stage varies among cultivars and is influenced by environmental factors. Stage I includes seed germination, emergence, and early vegetative growth. This stage is initiated at planting and terminates after the third trifoliolate leaf opens. Stage II includes the period between emergence of the third trifoliolate leaf and opening of the first flower. This period is characterized by branching and rapid vegetative growth. A new trifoliolate leaf develops on the main stem approximately every 3 days under favorable growing conditions. Stage III is the period between the first flower and mid pod set. At the end of stage III, floral initiation has normally ceased and 50% of the pods are fully elongated. Stage IV starts when the first formed pods begin to fill and continues until harvest maturation. Near the end of stage IV, small pods that have not started to fill cease to develop,

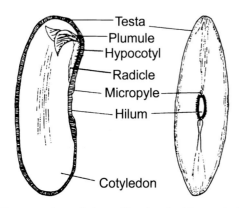

Fig. 1. Bean seed morphology. (Reprinted, with permission from Schwartz, Brick, Harveson, and Franc, 2004)

1) Seed
2) Radicle
3) Hypocotyl
4) Hypocotyl Arch
5) Tap Root
6) Lateral Roots
7) Root Hairs
8) Nodules
9) Cotyledons
10) Unifoliolate Leaves
11) Terminal Buds
12) Axillary Buds
13) Trifoliolate Leaves

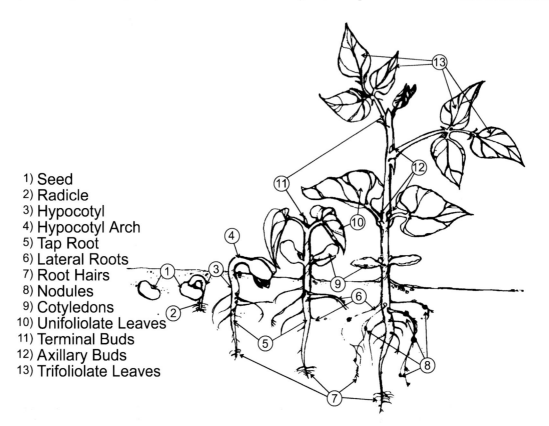

Fig. 2. Seed germination and seedling development. (Reprinted, with permission from Schwartz, Brick, Harveson, and Franc, 2004)

while existing pods and leaves begin to senesce. Upon completion of stage IV, the plant is mature and ready for harvest.

Considerable variation exists in the growth characteristics of common bean plants. This variation is used to separate germ plasm into four classifications based on determinacy of growth and plant architecture. The classifications include type I (determinate, bush), type II (indeterminate, upright), type III (indeterminate, semivine), and type IV (indeterminate, climbing vine) (Table 2). Classes are also subdivided into an "a" or "b"

subclassification based on the presence of a guide. Type I plants have a determinate growth habit, distinguished by a thick main stem, reproductive terminal buds, few internodes, short floral duration period (12–21 days), and more-uniform pod maturity than that of the other types. Because type I plants usually have uniform pod maturity, most snap bean cultivars developed for mechanical harvest have type I growth habit. Type Ia plants have a strong main stem and upright branches, whereas type Ib plants have weak branches and main stem and possess some

I. EMERGENCE AND EARLY VEGETATIVE GROWTH

The hypocotyl emerges from the soil (crook stage). — VE

The two cotyledons visible above ground at node 1. — VC
The two primary leaves (unifoliolate) unfolded at node 2.

The first trifoliolate leaf unfolded at node 3. — V1

The second trifoliolate leaf unfolded at node 4. — V2

The third trifoliolate leaf unfolded at node 5. — V3

II. BRANCHING AND RAPID VEGETATIVE GROWTH

The fourth trifoliolate leaf unfolded at node 6. — V4

Branches develop in the leaf axes and rapid growth occurs as new nodes develop on the main stem and/or branches every 3 to 5 days.

The $(n)^{th}$ trifoliolate leaf unfolded at node (n+2). — Vn

(Vegetative)

III. FLOWERING AND POD FORMATION

One open flower (early flower). — R1

50% open flowers (mid flower). — R2

One pod has reached maximum length (early pod set). — R3

50% of pods have reached maximum length (mid pod set). — R4

IV. POD FILL AND MATURATION

One pod with fully developed seeds (early seed fill). — R5

50% of pods with fully developed seeds (mid seed fill). — R6

One pod has changed from green to mature color such as striped, yellow, tan, purple (physiological maturity). — R7

80% of pods have changed to mature color (harvest maturity). — R8

(Reproductive)

Fig. 3. The four major growth and developmental stages in the vegetative and reproductive development of determinate and indeterminate bean plants. For determinate (type I/bush) beans, stems and branches terminate in an inflorescence. For indeterminate (type II and type III/vine) beans, stems and branch terminals remain vegetative. (Courtesy H. F. Schwartz and M. S. McMillan)

ability to climb. Type II and type III bean plants have an indeterminate growth habit distinguished by vegetative terminal buds on the main stem and lateral branches. This allows the plant to continue vegetative growth during flowering and pod fill. Type II and type III plants produce more nodes and usually have a longer floral duration than do type I plants. Type II plants are distinguished from type III plants by their erect main stem and branches, whereas type III plants have a weak main stem that grows semiprostrate or is twining to produce a dense canopy. Type IIa and type IIIa plants lack a strong terminal guide or leader and thus lack climbing ability, while type IIb and type IIIb plants have a terminal guide of varying length and possess some climbing ability. Type IV plants have very weak and excessively long stems and branches. They possess strong climbing ability and a structural support is necessary for maximum production. Type IVa plants have the pod load distributed all along the length of the plant, whereas type IVb plants have pods borne mostly on the upper part of the plant. Because the environment influences the number, length, and erectness of branches and the strength of the main stem, some cultivars vary in architecture across environments, whereas some cultivars are stable in their classification. Therefore, the characterization of growth habit for some genotypes may only be useful in a given environment.

Current breeding programs use broad-based parental germ plasm to improve disease resistance and agronomic characteristics. The categorization of *P. vulgaris* germ plasm into six races within two primary centers of domestication has contributed immensely to our knowledge about genetic diversity in *Phaseolus* germ plasm. The primary centers of domestication are Andean and Middle American. The Andean center of diversity from South America is further subdivided into three races that include Nueva Granada (northern Andes), Peruvian (Peruvian highlands), and Chilean (northern Chile and Argentina). Race Nueva Granada is represented by large-seeded light and dark red kidney bean, cranberry, yellow or Azufrado, and Calima beans, as well as other mottled types known as sugar and speckled beans, which are grown widely in Africa and the Caribbean. Germ plasm of races Peruvian and Chilean are not widely grown commercially in the Northern Hemisphere but include Nuña (popping beans) and coroscos. Beans from the Middle American center of diversity were domesticated in Mexico and Central America and include the races Durango (central highlands of Mexico), Jalisco (coastal Mexico near the state of Jalisco), and Mesoamerica (lowland tropical Central America). Market classes that typify race Durango include pinto, great northern, small red, pink, and bayo. Race Jalisco is composed primarily of photoperiod-sensitive material, such as Flor de Mayo and Flor de Junio. Race Mesoamerica is represented by small-seed classes, such as navy, small white, carioca, and black beans. Based on phaseolin type, snap beans are considered to be derived from the Andean center of domestication. However, some snap bean cultivars possess genetic material from both centers of domestication based on molecular markers. This categorization of centers of domestication and races has provided us with a better understanding of combining traits among races, coevolution of genes between the host and pathogen, and genetic incompatibility factors between genes from different centers of origin that can result in lethality of F_1 hybrids, give rise to crippling of trifoliolate leaves, or both.

In summary, common beans possess an immense spectrum of genetic diversity for morphological, architectural, nutritional, and economically important traits that have been exploited for human benefit throughout the world. Our understanding of interactions between host genes and pathogens continues to be an important component of bean research. New information about host–pathogen interactions will continue to enhance production and production efficiency in the future. We must continue our quest for a better understanding of both the host and pathogens to ensure that we can build upon our past accomplishments.

TABLE 1. Developmental Stages of Common Bean

Stage	Description
V1	*Emergence*: from the appearance of cotyledons on the soil surface to the unfolding of the primary leaves
V2	*Primary leaves*: from the full unfolding of the primary leaves to the unfolding of the first trifoliolate leaf
V3	*First trifoliolate leaf*: from the full unfolding of the first trifoliolate leaf to the unfolding of the third trifoliolate leaf
V4	*Third trifoliolate leaf*: from the full unfolding of the third trifoliolate leaf to the appearance of the first floral bud or raceme
R5	*Preflowering*: from the appearance of the first floral bud or raceme to the opening of the first flower
R6	*Flowering*: from the opening of the first flower to the expansion of the ovary after fertilization
R7	*Pod development*: from the expansion of the ovary to the elongation of the pod to its full size before increase in seed weight
R8	*Pod filling*: from the beginning of seed weight and size increase to the development of pigmentation of seeds and onset of leaf senescence
R9	*Maturity*: from initiation of senescence to complete senescence and drop in seed moisture to about 15%

TABLE 2. Growth Habit Classification and Description of Common Bean

Growth Habit	Description[a]
Type I	*Habit determinate Terminal bud reproductive Stem and branches erect or prostrate Terminal guide absent or small Pods distributed along the length of the stem
Type II	*Habit indeterminate Terminal bud vegetative *Stem and branches erect Terminal guide absent or medium Pods distributed along the length of the stem
Type III	Habit indeterminate Terminal bud vegetative *Stem and branches prostrate with little or no climbing ability Terminal guide small or long *Pods distributed mainly in the basal portion
Type IV	Habit indeterminate Terminal bud vegetative Stem and branches twining with strong climbing ability Terminal guide long or very long *Pods distributed along the length of the stem or mainly in the upper portion

[a] Key characteristics of the growth habit classification are marked with an asterisk.

Selected References

Brick, M. A., and Johnson, J. J. 2004. Classification, development and varietal performance. Pages 7-13 in: Dry Bean Production and Pest Management, 2nd ed. H. F. Schwartz, M. A. Brick, R. M. Harveson, and G. D. Franc, eds. Reg. Bull. 562A. Colorado State University, Fort Collins, CO.

Debouck, D. 1991. Systematics and morphology. Pages 55-118 in: Common Beans: Research for Crop Improvement. A. van Schoonhoven and O. Voysest, eds. CAB International, Wallingford, U.K., and CIAT, Cali, Colombia.

Debouck, D. 1999. Diversity in *Phaseolus* species in relation to the common bean. Pages 25-52 in: Common Bean Improvement in the

Twenty-First Century. S. P. Singh, ed. Kluwer Academic Publishers, Dordrecht, the Netherlands.

Debouck, D., and Smartt, A. J. 1995. Beans, *Phaseolus* spp. (Leguminosae-Papilionoideae). Pages 287-294 in: Evolution of Crop Plants, 2nd ed. J. Smartt and N. W. Simmonds, eds. Longman, London, U.K.

Gepts, P. 1998. Origin and evolution of common bean: Past events and recent trends. HortScience 33:1124-1130.

Gepts, P., and Debouck, D. 1991. Origin, domestication, and evolution of the common bean (*Phaseolus vulgaris* L.). Pages 7-53 in: Common Beans: Research for Crop Improvement. A. van Schoonhoven and O. Voysest, eds. CAB International, Wallingford, U.K., and CIAT, Cali, Colombia.

Kaplan, L. 1967. Archeological *Phaseolus* from Tehuacán. Pages 201-211 in: The Prehistory of the Tehuacán Valley, Vol. 1: Environment and Subsistence. D. E. Beyers, ed. University of Texas, Austin.

Kaplan, L. 1980. Variation in the cultivated beans. Pages 145-148 in: Guitarrero Cave: Early Man in the Andes. T. F. Lynch, ed. Academic Press, New York.

Kaplan, L., and McNesh, R. S. 1960. Prehistoric bean remains from caves in the Ocampo region of Tamaulipas, Mexico. Bot. Mus. Leafl. Harv. Univ. 19:33-56.

Koinange, E. M. K., Singh, S. P., and Gepts, P. 1996. Genetic control of the domestication syndrome in common-bean. Crop Sci. 36:1037-1045.

Myers, J. R., and, Baggett, J. R. 1999. Improvement of snap beans. Pages 289-330 in: Common Bean Improvement in the Twenty-First Century. S. P. Singh, ed. Kluwer Academic Publishers, Dordrecht, the Netherlands.

Schwartz, H. F., Brick, M. A., Harveson, R. M., and Franc, G. D., eds. 2004. Dry Bean Production and Pest Management, 2nd ed. Reg. Bull. 562A. Colorado State University, Fort Collins, CO.

Singh, S. P. 1982. A key for identification of different growth habits of *Phaseolus vulgaris* L. Annu. Rep. Bean Improv. Coop. 25:92-94.

Singh, S. P. 1999. Production and utilization. Pages 1-24 in: Common Bean Improvement in the Twenty-First Century. S. P. Singh, ed. Kluwer Academic Publishers, Dordrecht, the Netherlands.

Singh, S. P., and Gutiérrez, J. A. 1984. Geographical distribution of the DL_1 and DL_2 genes causing hybrid dwarfism in *Phaseolus vulgaris* L., their association with seed size, and their significance to breeding. Euphytica 33:337-345.

Singh, S. P., and Molina, A. 1996. Inheritance of crippled trifoliolate leaves occurring in intensive crosses of common bean and its relationship with hybrid dwarfism. Euphytica 26:665-679.

Singh, S. P., Gepts, P., and Debouck, D. G. 1991. Races of common bean (*Phaseolus vulgaris* L.). Econ. Bot. 45:379-396.

Singh, S. P., Gutiérrez, J. A., Molina, A., Urrea, C., and Gepts, P. 1991. Genetic diversity in cultivated common bean: II. Marker-based analysis of morphological and agronomic traits. Crop Sci. 31:23-29.

Skroch, P. W., and Nienhuis, J. 1995. Qualitative and quantitative characterization of RAPD variation among snap bean (*Phaseolus vulgaris*) genotypes. Theor. Appl. Genet. 91:1078-1085.

Voysest, O., Valencia, M. C., and Amezquita, M. C. 1994. Genetic diversity among Latin American Andean and Mesoamerican common bean cultivars. Crop Sci. 34:1100-1110.

White, J. F., and Laing, D. R. 1989. Photoperiod response of flowering in diverse genotypes of common bean (*P. vulgaris*). Field Crops Res. 22:113-128.

(Prepared by T. E. Michaels; Revised by M. A. Brick)

diseases, such as bean golden mosaic and southern blight, are important in localized geographic areas. Still other diseases, such as wildfire, are of minor importance. Annual production losses in world bean production as a result of diseases average about 10%. In view of the large number of potential diseases, their wide distribution, and the capacity of several diseases to cause extensive crop damage, it is apparent that losses in bean production would be much higher in the absence of disease management practices. A concerted, worldwide approach to the study of bean diseases and their management ensures the continued production of this important food legume for direct human consumption.

A plant is diseased when it is not functioning normally. Disease is the result of an interaction among the plant, its environment, and one or more harmful factors in the environment. These harmful agents may be infectious organisms (such as fungi, bacteria, nematodes, and phytoplasmas) or infectious agents (such as viruses, viroids, and related entities) that can reproduce only in the living plant. Plant diseases are also caused by abiotic agents, such as toxic chemicals, nutrient deficiencies, drought, and heat. Biotic and abiotic agents that cause disease are called pathogens. The visible indications of distress shown by diseased plants are called symptoms and may include yellowing (chlorosis) of leaves, discolorations, dead spots or patches (necrosis), wilting, stunting, malformations, and numerous other irregularities. The abnormal functioning of the plant generally leads to reductions in quantity and quality of harvested pods or seeds. Parts of the pathogen seen on or in diseased plants are called signs of the disease; examples include fungal mycelium, masses of white to colored spores, and brown to black sclerotia. Symptoms and signs are very useful in determining the cause of a disease. Accurate disease diagnosis is critical to developing and recommending effective disease management procedures.

Selected References

Agrios, G. N. 2005. Plant Pathology, 5th ed. Academic Press, San Diego, CA.

Allen, D. J., and Lenné, J. M. 1998. The Pathology of Food and Pasture Legumes. CAB International, Wallingford, U.K.

Schwartz, H. F., and Pastor-Corrales, M. A., eds. 1989. Bean Production Problems in the Tropics, 2nd ed. Centro Internacional de Agricultura Tropical (CIAT), Cali, Colombia.

Schwartz, H. F., Brick, M. A., Harveson, R. M., and Franc, G. D., eds. 2004. Dry Bean Production and Pest Management, 2nd ed. Reg. Bull. 562A. Colorado State University, Fort Collins, CO.

Wortmann, C. S., Kirkby, R. A., Eledu, C. A., and Allen, D. J. 1998. Bean diseases. Pages 63-86 in: Atlas of Common Bean (*Phaseolus vulgaris* L.) Production in Africa. CIAT Pub. No. 297. Pan-Africa Bean Research Alliance Report, Centro Internacional de Agricultura Tropical (CIAT), Cali, Colombia.

Zaumeyer, W. J., and Thomas, H. R. 1957. A Monographic Study of Bean Diseases and Methods for Their Control. U.S. Dep. Agric. Tech. Bull. 868.

(Prepared by R. Hall; Revised by H. F. Schwartz and R. Hall)

Bean Diseases

This compendium describes 73 bean diseases, 32 of which are caused by fungi, 5 by bacteria, 6 by nematodes, 26 by viruses, and 4 by phytoplasmas, plus numerous other noninfectious (abiotic) diseases and disorders. In addition, it describes the damage caused to bean plants by a wide range of environmental stresses. Some diseases, such as anthracnose, bean common mosaic, common bacterial blight, and white mold, can cause extensive or complete crop failure and are important throughout the bean production areas of the world. Other

Bean Pathogens

Most bean diseases are caused by fungi. Fungi that cause diseases in beans are microscopic organisms whose body cells resemble threads (called hyphae), which, en masse, form a mycelium. Hyphae may have cross-walls, called septa, and feed on the nutrients in the plant. Most fungi reproduce by forming specialized cells, called spores, that serve the fungus in many important ways. Some tolerate adverse conditions and permit the fungus to survive in the absence of the bean plant. Spores are an important means of dispersal and are often moved to the bean plant by wind, soil, water, or other agents. Once at a site suitable for infection under favorable environmental conditions, they germinate to produce germ tubes and, subsequently, new hyphae that penetrate the bean plant through wounds, natural openings, or the intact surface. Spores may also differ genetically from one another and thus enable new forms of the fungus to develop. Fungal pathogens of beans are identified mostly by the size, shape, and color of their spores. These and other structures, such as sclerotia and sexual fruiting bodies produced by fungal pathogens of beans, are described more fully under each pathogen. Fungal pathogens cause a wide range of symptoms on beans. Most frequently they cause variously colored (brown, yellow, red, or black) spots or blotches on leaves, stems, pods, seeds, or roots.

Bacteria that cause bean diseases are microscopic and appear cream colored or yellow en masse. They are rod shaped, motile, and generally gram negative; they do not form spores. They survive well in infected plant material, such as seeds and debris. However, in natural environments, they generally survive only for short periods apart from living plants or plant residues. They are dispersed by water, soil, infected plant parts, aerosols, and insects. They enter bean plants through natural openings or wounds and cause water-soaked spots and blotches that soon die and turn reddish brown on leaves, pods, or seeds.

Nematodes (eelworms) are microscopic, slender, wormlike animals. They move by swimming in films of water between soil particles or on plant surfaces. Nematodes develop from eggs, and the subsequent juvenile stages feed on the root system by puncturing the plant with a hollow, needlelike mouthpiece (stylet) and absorbing the plant cell contents. They are disseminated in water, soil, and plant material. In beans, they cause rotting or swelling (galls) of roots and may produce stunted plants.

Viral pathogens of beans are large, complex molecules composed of a nucleic acid core (either ribonucleic acid [RNA] or deoxyribonucleic acid [DNA]) and a protein coat. They can be seen with an electron microscope but are too small to be seen with a light microscope. Virus particles (virions) may be short or long rods or polyhedral in shape. They reproduce only in the living plant cells. They may be transmitted among plants in sap (mechanical transmission), in seeds, or by insects, such as aphids, whiteflies, and beetles. Common symptoms of viral infections include leaf mosaics (light and dark green or yellow areas), chlorosis, malformations (twisting or puckering), and plant stunting.

Phytoplasma organisms are prokaryotes and range in size from 175 nm to 150 μm. They are various shapes, including spherical, ovoid, and filamentous. They lack a cell wall; are bounded by a three-layered membrane; and contain cytoplasm, ribosomes, RNA, and DNA. They occur in phloem sieve tubes and are transmitted by leafhoppers. Symptoms of infection include yellowing, stunting, reddening, witches'-broom, and dieback.

Selected References

Agrios, G. N. 2005. Plant Pathology, 5th ed. Academic Press, San Diego, CA.
Allen, D. J., and Lenné, J. M. 1998. The Pathology of Food and Pasture Legumes. CAB International, Wallingford, U.K.
Schwartz, H. F., and Pastor-Corrales, M. A., eds. 1989. Bean Production Problems in the Tropics, 2nd ed. Centro Internacional de Agricultura Tropical (CIAT), Cali, Colombia.
Schwartz, H. F., Brick, M. A., Harveson, R. M., and Franc, G. D., eds. 2004. Dry Bean Production and Pest Management, 2nd ed. Reg. Bull. 562A. Colorado State University, Fort Collins, CO.
Zaumeyer, W. J., and Thomas, H. R. 1957. A Monographic Study of Bean Diseases and Methods for Their Control. U.S. Dep. Agric. Tech. Bull. 868.

(Prepared by R. Hall; Revised by H. F. Schwartz and R. Hall)

Bean Disease Management

Rational disease management recommendations are developed from detailed knowledge of the biology of the plant pathogen and the epidemiology of the disease. Thus, for each disease caused by an infectious organism or agent, this compendium describes survival, transmission, infection, host range, response to environment, and variability of the pathogen; genetic resistance and other characteristics of the plant; and cultural practices of the production system. From this knowledge, it is possible to develop effective scouting calendars and management strategies that reduce the harmful effects of the pathogen on the bean plant and that are consistent with economic production of the crop and protection of the environment. The most common management approaches for bean diseases include the use of disease-resistant cultivars, pathogen-free seeds, and cultural practices (e.g., crop rotation, crop residue management, and tillage practices) that suppress the pathogen or restrict its ability to spread or infect the plant. Another common avenue of disease management is the treatment of the soil, seeds, or crops with timely applications of chemical pesticides and biopesticides. The most effective and sustainable management of bean disease is obtained when several disease management methods are integrated with each other and with bean production practices.

Selected References

Agrios, G. N. 2005. Plant Pathology, 5th ed. Academic Press, San Diego, CA.
Allen, D. J., and Lenné, J. M. 1998. The Pathology of Food and Pasture Legumes. CAB International, Wallingford, U.K.
Schwartz, H. F., and Pastor-Corrales, M. A., eds. 1989. Bean Production Problems in the Tropics, 2nd ed. Centro Internacional de Agricultura Tropical (CIAT), Cali, Colombia.
Schwartz, H. F., Brick, M. A., Harveson, R. M., and Franc, G. D., eds. 2004. Dry Bean Production and Pest Management, 2nd ed. Reg. Bull. 562A. Colorado State University, Fort Collins, CO.
Zaumeyer, W. J., and Thomas, H. R. 1957. A Monographic Study of Bean Diseases and Methods for Their Control. U.S. Dep. Agric. Tech. Bull. 868.

(Prepared by R. Hall; Revised by H. F. Schwartz and R. Hall)

Bean Diagnostic Guidelines

Effective diagnostic procedure follows a logical sequence of steps, and considerable experience is necessary to achieve a high degree of competence in this process. This section provides guidelines regarding the sequential nature of disease diagnosis that may be helpful to less-experienced individuals. It is important to remember that multiple pathogens are commonly associated in a diseased plant and that disease symptoms vary greatly on different bean cultivars and under different environmental conditions. Classical disease symptoms and the abbreviated identification key provided here might, therefore, be a misleading oversimplification of the problem at hand.

Disease in beans, as in other plants, can be considered an interaction among susceptible bean genera and species (i.e., host or susceptible), the causal agent (pathogen), and an environment favorable for pathogen attack on the host. These contributing factors provide helpful sources of information needed for disease diagnosis.

Observe the Symptoms

The appearance of disease symptoms on affected beans is an important source of diagnostic information. However, one should view symptoms in the entire field, as well on individual bean plants.

Field symptoms are the visible disease patterns within the field. Stand symptoms are quite varied and are extremely important in disease diagnosis. For example, affected plants may be uniformly scattered across the area or they may be confined to low-lying wet areas or drier ridge tops. Disease may appear in the field as small circular spots, irregular concentrated patches, or large rings or circles, or it may have an unpatterned appearance or be uniform throughout the field. Occasionally, injury appears in the affected fields as bands, streaks, or other patterns that suggest mechanical damage from farm equipment or misapplication of fertilizers or pesticides. The soil under the plant and the roots should be closely inspected for insect pests or evidence of excessively applied fertilizers or other granular chemicals.

Individual plant symptoms are generally more apparent than are overall crop symptoms and include leaf spots, stem or petiole lesions, leaf blighting, wilting, yellowing, mottling, stunting, necrosis, and root rots. It is essential to observe carefully what parts of individual plants are affected. Leaf spots can be very diagnostic, since leaf spots are often distinctive in shape, color, and size for an individual disease. Leaf blighting, wherein the tissue dies quickly, is a symptom that is distinct from leaf spots. Leaf-blighting fungi typically produce a brownish necrosis or rot of leaf tissue, which lacks a definite form that is characteristic of leaf spots. Blighted areas in leaves can be any size or shape and may involve sectors between leaf veins of the entire leaf.

Symptoms of viral infection differ according to the cultivar of bean and virus species. Bean cultivars that are susceptible to viruses may respond by turning yellow or forming necrotic local lesions, mild systemic mottling, or severe mosaic and stunting, or they may remain symptomless.

Symptoms on pods may appear as dark green, irregularly shaped, blotched areas on green-podded types and as greenish yellow areas on yellow-podded types. Pod deformation with poor seed set and reduced seed size or plants that are completely barren of flowers and pods may also occur.

Observe the Environmental Conditions

The cultural and climatological environment prior to and during the onset of disease problems is an additional source of diagnostic information. Temperature, light intensity and duration, and moisture conditions provide critical diagnostic information. The nature of the affected site is also important. Air, water drainage, soil type, sun exposure, terrain slope, age of the stand, and proximity of windbreaks or structures may all be significant factors in bean disease development. Prior chemical applications (including pesticides, fertilizers, and industrial waste from a leak or spill [i.e., lubricating oil]) to the site may also provide important clues in disease diagnosis. Dense bean stands may trigger or amplify certain disease problems by altering the microenvironment.

Identify Visible or Microscopic Signs of the Pathogen

The appearance of the causal agent, when it can be observed, is a critical and important aid to disease diagnosis. This is true whether the observation of the pathogen is macroscopic, microscopic, or both. Certain structures of some bean pathogens may be visible without magnification. For diseases such as powdery mildew and rust, black, white, or orange spores of the causal fungus often are visible on affected bean foliage. Observation of lower leaves may reveal other signs, such as infection mats, acervuli, or sclerotia. When the causal fungus is visible, its appearance is often the most important clue in disease identification.

The use of a 10–20× hand lens may assist in identifying the cause of observed abnormalities. Mechanical and insect feeding injury may be differentiated from leaf damage caused by infectious agents under the lens. Fungal fruiting bodies, such as

acervuli, sporodochia, pycnidia, or perithecia, may be visible with low magnification.

If microscopic examination of affected plant parts does not reveal the presence of fungal structures, tissues showing symptoms may be incubated for 24–48 h in a moist chamber (e.g., plastic bag with wet paper towel at 25°C). The moist environment may induce sporulation or hyphal proliferation. It must be remembered, however, that any fungus present on infected tissue, including saprophytes and epiphytes, may grow in such an environment. When spores or other structures are produced on infected tissue, this material can be mounted in tap water on a glass slide for microscopic examination.

Small, dark structures that are sometimes present in leaf or stem lesions may be the necks of flask-shaped pycnidia produced by fungi in the subdivision Deuteromycotina (order Sphaeropsidales) or the perithecia of fungi in the subdivision Ascomycotina. These structures can be removed from the leaf or stem tissue with a needle, mounted intact in a drop of water on a glass slide, and gently crushed by applying pressure to a glass coverslip placed on the drop of water. Mature spores will ooze out of the pycnidia or perithecia and may be identifiable.

If small, spore-bearing fruiting bodies or cushions (acervuli or pycnidia) are present in leaf lesions, it may be necessary to cut thin cross sections of these structures to observe the morphology of the fruiting body. This technique is particularly useful for fungi that produce spores in acervuli (*Colletotrichum* spp.) or pycnidia (*Ascochyta* or *Phoma* spp.).

There are three blights of beans caused by bacteria: common blight, caused by *Xanthomonas axonopodis* pv. *phaseoli* (Smith) Vauterin et al.; halo blight, caused by *Pseudomonas syringae* pv. *phaseolicola* (Burkholder) Young et al.; and bacterial brown spot, caused by *Pseudomonas syringae* pv. *syringae* van Hall. The symptoms appear first on the lower side of the leaves as small, water-soaked spots and many have yellow halos. The spots enlarge, coalesce, and may form large areas that later become necrotic. When small slices of the infected leaves or pods are placed on a slide in a drop of water and observed with a dissecting scope, a cloud of bacterial ooze may be evident emerging from the cut edge.

If pathogenic fungi and bacteria are not apparent on or within affected tissue, symptoms may have resulted from viral infection or abiotic causes, such as excessive fertilizers, salts, herbicides, heat, moisture, cold, atmospheric pollution, soil pH, mechanical injuries, or other noninfectious agents.

The foregoing techniques are particularly useful in attempts to diagnose foliar diseases of beans. Because soils abound with saprophytic fungi, it may be more difficult to determine the cause of injury or disease occurring in roots and belowground stems. Diagnostic techniques may initially include washing the roots and preparing thin sections of symptomatic tissue to view with a microscope. Other techniques include attempts to isolate causal pathogens on various culture media, treating root or stem tissues with chemicals to render them transparent, applying stains to selectively color fungal structures, and growing susceptible plants over infected tissue to trap a pathogen. These techniques require specialized equipment and considerable skill on the part of the diagnostician.

Collection and Submission of Samples

Samples for disease diagnosis should be taken when the problem is active or increasing. Areas should be selected for sampling where the damage or symptoms are representative of the entire affected area, and samples should be obtained before the application of pesticides. Samples should be collected at the edge of an infected area, and they should include both healthy and infected plants exhibiting various stages of infection. Samples taken weeks after the onset of symptoms or when plants are senescing are generally of little value to the diagnostician.

As soon as possible after collection, samples should be wrapped in dry paper toweling (wet paper toweling frequently leads to sample decay prior to arrival at the clinic); placed in a container, such as a self-sealing plastic bag, to prevent desiccation; and delivered promptly to the clinic. In cases in which the whole plant is collected, it is helpful to wrap the root system in a smaller plastic bag to prevent contamination of foliage with soil particles during handling and shipment.

Since the clinician who receives the sample will not have an opportunity to observe the problem in the field, it is essential that the following information accompany the sample: 1) the bean cultivar affected and extent of damage; 2) the cultural and climatological environment of the affected area; 3) a complete, accurate description of the symptoms and date of first appearance on individual plants (Polaroid or digital photographs of symptoms included with samples are very useful to diagnosticians); and 4) the chemicals (e.g., herbicides) used on or near the crop during that season or the previous season.

For the most accurate diagnosis, samples should reach the diagnostician in approximately the condition they were in when collected. Rapid delivery of samples to the clinic is essential. If personal delivery of the sample is not possible, the quickest carrier available should be used. Priority mail or overnight delivery services are valuable for this purpose. Samples should never be sent at the end of the week, since this may result in the sample sitting undelivered over the weekend.

(Prepared by R. T. McMillan)

Bean Crop Health and Integrated Pest Management

The major components of pest management that provide the foundation for integrated pest management (IPM) consist of exclusion, eradication, protection, resistance, and biological management.

Exclusion prevents the entrance and establishment of a plant pathogen in an uninfested region by prohibition, interception, and elimination. This is primarily accomplished by implementation of quarantine, embargo, inspection, and certification of plant materials, such as bean seeds, that are transported between production regions.

Eradication emphasizes the removal, elimination, or destruction of a pest from an area in which it is already established. This is primarily accomplished for beans by detection and removal of diseased plants or infested plant debris; elimination of weed hosts and alternate crops; sanitation; crop rotation; and destruction of pests by disinfestation or disinfection by heat (burning), tillage, or pesticides.

Protection is the placement of a protective barrier between the susceptible host and the pest. Application of protectant or systemic chemicals, such as fungicides, bactericides, insecticides, nematicides, and herbicides, is the best example of this principle, but manipulation of environmental factors, usually by cultural practices such as planting time, row orientation, and row spacing/plant density, are also beneficial.

Resistance to a pathogen is genetically inherited and is usually expressed as a result of an interaction between a pest and the host at the cellular level that limits, but does not eliminate, pest development; that elicits plant defense mechanisms; or both. Resistance also helps plants escape disease (via modifications in canopy architecture, flowering, and maturity). Tolerance, whereby plants are susceptible to disease but yield and quality are not diminished, is often mistakenly referred to as resistance. Resistance strategies used with beans include nutrition to increase plant vigor but not to overstimulate vegetative growth; selection of plants that are less affected by a pest, and multiplication of seeds of the selected plants; and hybridization and selection of desirable progeny from crosses between

susceptible and less-susceptible or resistant parents. The developing field of genetic engineering offers much promise for improved resistance, particularly in those species in which limited progress has been made to date.

Biological management is the use of natural enemies and competitors to manage insect pests, pathogens, and weeds. This may be accomplished through conservation of existing biological control agents via appropriate choice of selective pesticides to be applied, augmentation of biological control agent population densities through release of additional individuals, or inundation through the release of sufficient biological control agents to reduce the pest population densities to a subeconomic damage level. An additional approach, often termed classical biological management, is to collect natural enemies of an exotic pest within its native range and release them into areas where the pest has been introduced.

Bean Crop Health

This section contains excerpts from the review chapter on IPM published in the authoritative book edited by S. P. Singh entitled *Common Bean Improvement in the Twenty-First Century*. The goal of IPM is to achieve relief from pests in a manner that ensures safety, profitability, and durability while shifting the perceived focus on pesticide-intensive strategies to a systems approach that emphasizes biological knowledge of pests and their interactions with crops. The shift to a systems approach with primary and secondary tactics requires that we move toward management of all pests rather than target only key pests, such as a single insect, pathogen, or weed species. Primary tactics rely upon knowledge of the managed ecosystem and its natural processes that suppress pest populations; and secondary tactics rely upon intervention with pesticides or physical or biological supplementation. A simplistic categorization of bean IPM strategies during the last 25 years in temperate-cropping systems would show that the bean industry has emphasized pest management via pesticides as a primary tool, followed closely by plant resistance and, finally, by cultural practices. On the other hand, tropical-cropping systems have emphasized traditional cultural practices, followed by plant resistance, with pesticide management often assuming a distant third.

It is also important to note the important common role that host resistance plays across bean production systems. Host resistance is reviewed in great detail in other references and is obviously an important fundamental of, but is not an exclusive substitute for, any IPM strategy that is devised and implemented for the dynamic complex of pests faced by bean crops in various cropping systems throughout the world. Reliance on a single management strategy, e.g., host resistance, runs counter to the very core of the IPM philosophy.

More in-depth sources of information on specific biotic stress management approaches are available from numerous research papers and reviews, including those by Allen et al., Beebe and Pastor-Corrales, Hall, Hall and Nasser, Schwartz and Pastor-Corrales, Schwartz and Peairs, van Schoonhoven and Voysest, and Wortmann et al.

Bean IPM Tactics

Hall and Nasser published a thorough review of 33 specific disease management elements that are utilized or recommended individually and collectively to manage 50 bean diseases worldwide. A review of the elements utilized to manage diseases reveals that only a small number of those elements actually constitute the bulk of the current IPM strategies employed. The major emphasis for most diseases has been on 1) crop rotation of more than 2 years between bean crops; 2) timely application of specific pesticides that are effective against the target pest; 3) tillage practices that promote host plant development, reduce carryover of the pathogen, or both; 4) weed management to reduce the level of host plant stress from competition, microclimatic influences, and sources of pests (refugia); 5) multiple cropping with nonhosts and crop barriers to reduce pathogen spread; and 6) plant resistance to one or more plant pathogens.

Cultural practices, which include crop rotation, are recommended as a management strategy for 80% of the bean diseases; whereas use of resistant cultivars is recommended as a management practice for 44% of the more-common bean diseases (e.g., rust, common bacterial blight, and bean common mosaic). The effectiveness of individual IPM elements is quite varied and unique for specific types of diseases and underscores the complexity surrounding the design and implementation of disease management strategies.

Resistance to many plant pathogens may not be complete enough to protect the plant from significant loss in yield and quality, so other elements, which include cultural practices, are essential to supplement the partial gain realized by resistance. In addition, resistance may be less effective because of genotype × environmental interactions, and few cultivars are likely to possess long-lasting resistances to all elements (i.e., races) of a pathogen or pest complex, thereby requiring a more complete protection strategy for overall plant health.

The following list of cultural practices has been applied in various combinations to bean IPM strategies in temperate- and tropical-cropping systems alike. Continued emphasis upon adoption of the primary elements will provide the most immediate impact upon priority pathogens and diseases, and adoption of the secondary elements can further enhance the effectiveness of IPM strategies in specific cropping systems in temperate and tropical regions.

Primary IPM Elements
- crop rotation
- eliminating volunteer beans and infested debris
- clean fallow and burning
- pathogen-free seeds
- cultivar selection for adaptation and disease resistance
- soil temperatures of 18°C or higher at planting
- seeding dates
- seeding depth
- suppressing weeds and insect pests
- tillage (preplant and postemergence)
- timely harvest

Secondary IPM Elements
- acceptable pH
- adequate level of organic matter
- adequate and balanced soil fertility
- soil moisture levels near field capacity
- judiciously using surface or overhead irrigation
- appropriate growth habit, plant population, and plant spacing
- row orientation
- separating bean fields and using biological barriers
- roguing
- limiting movement of equipment and personnel within and between fields

Selected References

Allen, D. J., Buruchara, R. A., and Smithson, J. B. 1998. Diseases of common bean. Pages 179-265 in: The Pathology of Food and Pasture Legumes. D. J. Allen and J. M. Lenné, eds. CAB International, Wallingford, U.K.

Beebe, S. E., and Pastor-Corrales, M. A. 1991. Breeding for disease resistance. Pages 561-617 in: Common Beans—Research for Crop Improvement. A. van Schoonhoven and O. Voysest, eds. CAB International, Wallingford, U.K., and CIAT, Cali, Colombia.

Hall, R. 1995. Challenges and prospects of integrated pest management. Pages 1-19 in: Novel Approaches to Integrated Pest Management. R. Reuveni, ed. Lewis Publishers, Boca Raton, FL.

Hall, R., and Nasser, L. C. B. 1996. Practice and precept in cultural management of bean diseases. Can. J. Plant Pathol. 18:176-195.

Maloy, O. C. 1993. Plant Disease Control—Principles and Practice. J. Wiley & Sons, Inc., New York.

Schwartz, H. F., and Pastor-Corrales, M. A. 1989. Bean Production Problems in the Tropics. Centro Internacional de Agricultura Tropical (CIAT), Cali, Colombia.

Schwartz, H. F., and Peairs, F. B. 1999. Integrated pest management. Pages 371-388 in: Common Bean Improvement in the Twenty-First Century. S. P. Singh, ed. Kluwer Academic Publishers, Boston, MA.

Schwartz, H. F., Brick, M. A., Harveson, R. M., and Franc, G. D., eds. 2004. Dry Bean Production and Pest Management, 2nd ed. Reg. Bull. 562A. Colorado State University, Fort Collins, CO.

van Schoonhoven, A., and Voysest, O. 1991. Common Beans: Research for Crop Improvement. CAB International, Wallingford, U.K., and CIAT, Cali, Colombia.

Wortmann, C. S., Kirkby, R. A., Eledu, C. A., and Allen, D. J. 1998. Atlas of Common Bean (*Phaseolus vulgaris* L.) Production in Africa. CIAT Pub. No. 297. Pan-Africa Bean Research Alliance Report, Centro Internacional de Agricultura Tropical (CIAT), Cali, Colombia.

(Prepared by H. F. Schwartz, J. R. Steadman, and R. Hall)

Part I. Infectious Diseases

Fungal Diseases of Subterranean Parts

Aphanomyces Root and Hypocotyl Rot

Aphanomyces euteiches Drechs. was known to infect beans under greenhouse conditions as early as the 1960s, but it wasn't until 1979 that *A. euteiches* was isolated from beans in the field. That year, a strain designated *A. euteiches* f. sp. *phaseoli* W. F. Pfender & D. J. Hagedorn, capable of causing severe root and hypocotyl rot in beans in the field, was discovered in central Wisconsin. *A. eutiches* f. sp. *phaseoli* has also been reported in New York, and *Aphanomyces* spp. have also been reported to cause root rot in Australia. Since *Aphanomyces* spp. generally are associated with other pathogenic fungi, they may be obscured on culture media and go undetected.

Symptoms

The pathogen can infect plants from soon after emergence to late in the season. It does not cause seed rot or preemergence damping-off. Lesions on roots are initially yellow-brown and fairly firm. They rapidly coalesce to involve most of the roots, which become softer as the cortex is destroyed. The infected roots soon darken from the activity of secondary invaders.

Plants may be severely stunted. Typically, the pathogen grows up the hypocotyl to produce a lesion above the soil line (Figs. 4 and 5). This lesion is slightly water-soaked and gray-green in appearance at its leading edge, but it becomes brown as the necrosis develops. Seedlings may be killed as the lesions extend to the growing point of the plant, especially in mixed infections with *Pythium* spp. (Figs. 6 and 7).

Fig. 5. Snap beans from a field infested with *Pythium ultimum* and *Aphanomyces euteiches* f. sp. *phaseoli*, showing moderate (left) and mild (right) root rot. (Courtesy D. J. Hagedorn, from the files of W. F. Pfender)

Fig. 6. Snap beans grown at 28°C in soil that were (left to right) uninfested, infested with *Pythium ultimum* alone, infested with *Aphanomyces euteiches* f. sp. *phaseoli* alone, and infested with both pathogens. (Courtesy D. J. Hagedorn, from the files of W. F. Pfender)

Fig. 4. Snap beans grown in soil containing *Aphanomyces euteiches* f. sp. *phaseoli* (center and right). Check is on the left. (Courtesy D. J. Hagedorn, from the files of W. F. Pfender)

Causal Organism

A. euteiches has aseptate mycelium and can produce two kinds of spores. Thick-walled oospores are formed by sexual fusion of the oogonia and antheridia (Fig. 8). When they germinate, oospores form hyphae or sporangia. The sporangia produce asexual swimming spores (zoospores). Primary spores are extruded from the sporangia in single file and encyst at the mouth of the sporangium. Within hours, zoospores emerge from these cysts, swim for a brief time, encyst again, and then germinate to produce hyphae that infect host tissue.

A. euteiches f. sp. *phaseoli* can be differentiated from other strains of *A. euteiches* by slow growth at 30°C and a larger aplerotic zone.

The fungus is readily isolated from newly infected plants before the infected tissue is severely rotted. Isolation from the advancing margin of the hypocotyl lesion is often successful. The pathogen grows quickly from segments of surface-sterilized tissue placed on water agar or on a semiselective culture medium. *Aphanomyces* spp. can be recognized by their sparse, arachnoid growth habit on cornmeal agar and by the characteristic appearance of the hyphae under microscopic observation. The identity of a suspected isolate of the genus *Aphanomyces* can be confirmed by inducing it to form its characteristic sporangia and zoospore cysts. This can be accomplished by floating small agar plugs, cut from the margin of a colony grown on cornmeal agar, in a dilute solution of salts or in dilute, sterilized lake water.

Disease Cycle and Epidemiology

Oospores can persist in a dormant state in the soil for years. Sporangia and zoospores are produced in infected roots and presumably are active in secondary cycles of the disease, although this has not been demonstrated in nature. They are produced in infected cortical tissue within 2 weeks after infection and are released to the soil as roots decay. Oospores may be dispersed long distances in infested soil.

The host range of *A. euteiches* f. sp. *phaseoli* includes alfalfa, as well as snap beans and dry beans. The fungus does not cause severe disease on pea, soybean, or common hosts of several other *Aphanomyces* spp. Some *A. euteiches* isolates can produce disease on beans but cause much less damage than does *A. euteiches* f. sp. *phaseoli* (Fig. 9).

To date, *A. euteiches* f. sp. *phaseoli* has been reported only from irrigated, sandy soils. Whether it is limited to this environment is not known, but an unidentified *Aphanomyces* species was important in a root rot complex in a clay loam soil in Australia. When beans are grown without rotation on irrigated,

sandy soils, the population of *Aphanomyces* spp. can increase rapidly (during one to two seasons) to very damaging levels. Severe losses have occurred in the second year of production on land that had been previously uncultivated. The host range and persistence of this pathogen in weeds have not been determined.

Soil moisture and temperature are important determinants of disease severity. Since *Aphanomyces* spp. are water molds, they are most active at high soil moisture levels and cause the most severe disease during wet seasons and in fields irrigated frequently. Infection can occur at all temperatures favorable to bean growth, but disease is more severe at 20–28°C (Fig. 6) than at 16°C (Fig. 7).

Pythium ultimum Trow is frequently associated with *Aphanomyces* spp. in infected plants. In controlled studies with inoculum levels similar to those in field soil, *A. euteiches* f. sp. *phaseoli* was more damaging than was *P. ultimum* at 20–28°C but less damaging than was *P. ultimum* at 16°C. Mixed infections by the two pathogens increase disease severity synergistically and cause increased mortality of infected plants, especially at higher temperatures.

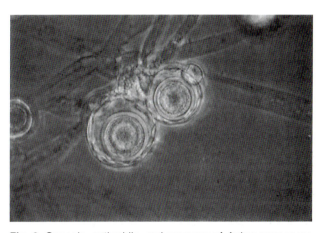

Fig. 8. Oogonia, antheridia, and oospores of *Aphanomyces euteiches* f. sp. *phaseoli*. (Courtesy W. F. Pfender)

Fig. 9. Snap beans infected with *Aphanomyces euteiches* f. sp. *pisi* W. F. Pfender & D. J. Hagedorn (left) and *A. euteiches* f. sp. *phaseoli* (right). Check is in the center. (Courtesy D. J. Hagedorn, from the files of W. F. Pfender)

Fig. 7. Snap beans grown at 16°C in soil that were (left to right) uninfested, infested with *Pythium ultimum* alone, infested with *Aphanomyces euteiches* f. sp. *phaseoli* alone, and infested with both pathogens. (Courtesy D. J. Hagedorn, from the files of W. F. Pfender)

Management

The most effective management for this disease is avoidance of infested fields. Soil from prospective bean fields should be tested in the laboratory for its root rot potential. Regular rotation of crops may be helpful in delaying buildup of *Aphanomyces* populations. Chemical management is not currently available.

Management through genetic resistance is being investigated, and resistance has been found in some breeding lines. However, genetic linkages between resistance and agronomically undesirable traits must be overcome. The incorporation of crucifer green manures has shown some potential for reducing the incidence of Aphanomyces root rot in several crops, including beans.

Selected References

Carley, H. E. 1970. Detection of *Aphanomyces euteiches* races using a differential bean series. Plant Dis. Rep. 54:943-945.

Delwiche, P. A., Grau, C. R., Holub, E. B., and Perry, J. B. 1987. Characterization of *Aphanomyces euteiches* isolates recovered from alfalfa in Wisconsin. Plant Dis. 71:155-161.

Holub, E. B., Grau, C. R., and Parke, J. L. 1991. Evaluation of the forma specialis concept in *Aphanomyces euteiches*. Mycol. Res. 95:147-157.

Malvick, D. K., Grau, C. R., and Percich, J. A. 1998. Characterization of *Aphanomyces euteiches* strains based on pathogenicity tests and random amplified polymorphic DNA analyses. Mycol. Res. 102:465-475.

O'Brien, R. G., O'Hare, P. J., and Glass, R. J. 1991. Cultural practices in the control of bean root rot. Aust. J. Exp. Agric. 31:551-555.

Pfender, W. F., and Hagedorn, D. J. 1982. *Aphanomyces euteiches* f. sp. *phaseoli*, a causal agent of bean root and hypocotyl rot. Phytopathology 72:306-310.

Pfender, W. F., and Hagedorn, D. J. 1982. Comparative virulence of *Aphanomyces euteiches* f. sp. *phaseoli* and *Pythium ultimum* on *Phaseolus vulgaris* at naturally occurring inoculum levels. Phytopathology 72:1200-1204.

Scott, W. W. 1961. A monograph of the genus *Aphanomyces*. Va. Agric. Exp. Stn. Tech. Bull. 151.

Stamps, D. J. 1978. *Aphanomyces euteiches*. Descriptions of Pathogenic Fungi and Bacteria, No. 600. Commonwealth Mycological Institute and Association of Applied Biologists, Kew, Surrey, England.

(Prepared by W. F. Pfender; Revised by L. E. Hanson)

Black Root Rot

Black root rot is caused by *Thielaviopsis basicola* (Berk. & Broome) Ferraris (syn. *Chalara elegans* Nag Raj & Kendrick). The pathogen is widely distributed, can infect more than 130 plant species in 15 families, and causes severe black root rot diseases in ornamentals and crops, such as bean, carrot, cotton, pea, peanut, tomato, and tobacco. Black root rot can be a problem on beans in limited areas of the United States, Italy, and Germany, but it does not appear to be of significance in Latin America.

Symptoms

Initial symptoms of black root rot in beans appear as elongated, narrow lesions on the hypocotyl and root tissues. These lesions are initially reddish purple and then become dark charcoal to black. As the disease progresses, the lesions often coalesce and form large black areas (Fig. 10). The lesions may remain superficial and cause limited damage or become deep and cause stunting, premature defoliation, and eventually plant death.

Causal Organism

T. basicola grows and sporulates readily on artificial agar media but exhibits considerable variation in colony appearance, zonation, growth rate, and shape and number of spores produced. The pathogen can be easily isolated from soil and infected tissues on fresh carrot disks or on culture media selective for the fungus. Asexual reproduction of *T. basicola* occurs by the formation of endoconidia (phialospores) and chlamydospores (aleuriospores). Endoconidia are hyaline, small, and cylindrical. They are produced within the conidiophores (phialides) and are extruded singly or in chains. Chlamydospores (Fig. 11) are thick walled, dark brown, and multicellular and are produced laterally or terminally in the hyphae. Cells of the chlamydospores eventually separate and germinate as individual infection units. Chlamydospores survive for long periods in soil, whereas endoconidia are short-lived. The pathogen is disseminated within and between bean fields by movement of infested soil, infected host tissues, colonized debris, drainage water, or irrigation water, as well as by other means.

Disease Cycle and Epidemiology

Chlamydospores of *T. basicola* in soil germinate to produce multiple germ tubes and (eventually) several hyphae, which grow toward and onto the hypocotyl and root surfaces. These hyphae penetrate bean tissues through wounds, lesions incited by other bean pathogens, or the intact surface (without forming appressoria). Phosphatidase enzymes may play a major role in the initial penetration of epidermal cells and in later phases of pathogenesis. After penetration, the pathogen produces constricted and nonconstricted hyphae that grow within and between plant cells, respectively. Chlamydospores are produced by nonconstricted hyphae throughout the infected tissues. Under moist conditions, reproductive hyphae emerge through the epidermal layer and eventually produce masses of chlamydospores and endoconidia. As infected host tissues decay, chlamydospores are

Fig. 10. Black root rot, caused by *Thielaviopsis basicola* (plants on the right). Healthy plant is on the left. (Courtesy D. J. Hagedorn, from the files of G. S. Abawi)

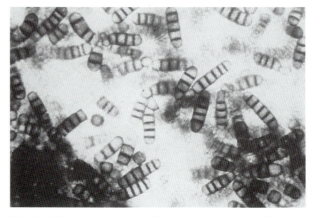

Fig. 11. Chlamydospores of *Thielaviopsis basicola*. (Courtesy G. S. Abawi)

released into the soil and, under favorable conditions, germinate to infect host tissue or colonize available organic debris.

T. basicola grows and sporulates most abundantly in the laboratory at relatively high temperatures (25–28°C) but damages beans most severely at lower temperatures (15–20°C). Black root rot is favored by wet, cool, neutral to alkaline soils and nitrogen fertilizers.

Management

Soil treatments with fungicides, such as thiabendazole and captan, or fumigants, such as methyl isothiocyanate and dazomet, are effective against *T. basicola* on beans. However, the use of such chemicals is very limited because they are expensive and difficult to apply.

Breeding lines have been produced with high levels of resistance to *T. basicola* and have been used in many breeding programs as sources of resistance to the black root rot pathogen. Three partially recessive genes appear to be responsible for this resistance through the production of two phytoalexins.

Lowering the soil pH to less than 5.2 can significantly reduce the incidence of black root rot but may cause reduced plant growth.

Selected References

Abawi, G. S. 1989. Root rots. Pages 105-157 in: Bean Production Problems in the Tropics, 2nd ed. H. F. Schwartz and M. A. Pastor-Corrales, eds. Centro Internacional de Agricultura Tropical (CIAT), Cali, Colombia.

Chittaranjan, S., and Punja, Z. K. 1994. Factors influencing survival of phialospores of *Chalara elegans* in organic soil. Plant Dis. 78:411-415.

Christou, T. 1962. Penetration and host-parasite relationships of *Thielaviopsis basicola* in the bean plant. Phytopathology 52:194-198.

Hassan, A. A., Wilkinson, R. E., and Wallace, D. H. 1971. Relationship between genes controlling resistance to Fusarium and Thielaviopsis root rots in bean. J. Am. Soc. Hortic. Sci. 96:631-632.

Papavizas, G. C., Lewis, J. A., and Adams, P. B. 1970. Survival of root-infecting fungi in soil. XIV: Effect of amendments and fungicides on bean root rot caused by *Thielaviopsis basicola*. Plant Dis. Rep. 54:114-118.

Pierre, R. E. 1971. Phytoalexin induction in beans resistant or susceptible to *Fusarium* and *Thielaviopsis*. Phytopathology 61:322-327.

Punja, Z. K., and Sun, L. J. 1999. Morphological and molecular characterization of *Chalara elegans* (*Thielaviopsis basicola*), cause of black root rot on diverse plant species. Can. J. Bot. 77:1801-1812.

Yarwood, C. E., and Levkina, L. M. 1976. Crops favoring *Thielaviopsis*. Plant Dis. Rep. 60:347-349.

(Prepared by G. S. Abawi; Revised by L. E. Hanson)

Fusarium Root Rot

Fusarium root rot (Fusarium foot rot, dry root rot) of beans occurs in most bean-growing areas throughout the world. The disease usually causes little damage in unstressed plants. However, under conditions of reduced root growth caused by drought, soil compaction, soil saturation after rill irrigation, or oxygen stress, Fusarium root rot can almost destroy a bean crop. Even the highest levels of resistance to the disease are overcome by the pathogen when fields are flooded or roots are deprived of oxygen for short time periods (e.g., 24 h). The disease is often found associated with Rhizoctonia root rot and Pythium root rot (and possibly others) in a complex (Fig. 12).

Symptoms

The first symptoms that appear are narrow, longitudinal, red to brown streaks on hypocotyls and taproots of 7- to 10-day-old seedlings. The cortex of the hypocotyl and older portions of the root become progressively more streaked and generally necrotic. Necrosis is largely confined to the cortex. If the root system grows unrestricted, plant productivity appears unaffected (Fig. 13).

Diseased plants often respond by producing numerous adventitious roots (Fig. 12) from the hypocotyl in longitudinal rows near the soil surface, particularly when soil is mounded around the base of the plant during cultivation. When the disease is severe, plants are stunted and typically depend on a clump of adventitious roots for survival. Primary leaves of such plants prematurely turn yellow and drop off, especially if plants are stressed by moisture extremes and soil compaction. Diseased plants vary in size and vigor within a field, creating an irregular crop canopy.

Causal Organism

The causal organism is *Fusarium solani* (Mart.) Sacc. f. sp. *phaseoli* (Burkholder) W. C. Snyder & H. N. Hans. The fungus produces septate, hyaline mycelium. Its macroconidia have mostly three septa (5.1 × 44.5 µm) and four septa (5.3 × 50.9

Fig. 12. Fusarium root rot, caused by *Fusarium solani* f. sp. *phaseoli*. Note proliferation of adventitious roots on the plant on the right following severe rotting of the lower roots. (Courtesy R. Hall)

Fig. 13. Improved root growth by subsoiling at a depth of 20 cm between rows (right) compared with lack of subsoiling (left). (Courtesy H. F. Schwartz)

μm), rarely five septa, of uniform diameter along their length, are curved, and are rounded or slightly pointed at the apex (Fig. 14). Microconidia are rare. Conidia are borne in sporodochia. Chlamydospores are globose (11.6 μm), terminal or intercalary in conidia or hyphae, and single or in short chains (Fig. 15). Cultures are relatively slow-growing compared with other formae speciales of *F. solani* and contain various shades of blue or green depending on the isolate and the culture medium (Fig. 16). No perfect state has been described for this forma specialis.

Disease Cycle and Epidemiology

The pathogen resides in soil mainly as thick-walled chlamydospores. These resting spores germinate readily when stimulated by nutrients (e.g., sugars and amino acids) exuded by germinating seeds and root tips. Resulting hyphae penetrate the bean plant directly and through stomata and wounds. Hyphae penetrate the intercellular spaces of the cortex but stop at the endodermis. Under conditions of high soil moisture levels, conidia may be produced on sporodochia emerging from stomata near the soil surface. As infected tissues degenerate, conidia and hyphae convert to chlamydospores to complete the life cycle.

Chlamydospores of *F. solani* f. sp. *phaseoli* can germinate and reproduce in soil near seeds and roots of many nonsusceptible plants and other organic matter. Thus, the fungus may survive in infested fields indefinitely, for example, in soil planted continuously to nonhost crops for more than 30 years.

F. solani f. sp. *phaseoli* is believed to disperse in soil, in dust, or on seeds. Infested soil and organic matter are also spread locally by wind and water. When beans are grown for the first time in fields adjacent to infested fields, however, the disease is

inconsequential until a third crop of beans is grown in that new field, when, for example, in sandy loam soils of the northwestern United States, the fungus becomes uniformly distributed in the plowed layer of soil. Observations suggest that *F. solani* f. sp. *phaseoli* can sustain itself as a saprophyte but reaches pathogenic potential only by multiplying on bean crops. Excellent methods have been developed to monitor this pathogen in soil. The disease is more severe in cool, moist soil, but suppression of yield is greater under drought conditions.

Management

Numerous efforts to reduce localized foot rot symptoms on the hypocotyl and taproot have failed. Rarely, if ever, have seed treatments or other localized chemical treatments increased bean yields. Bean yields in *Fusarium*-infested fields were not increased even by complete protection of hypocotyls and taproots if the remainder of the root system extended into infested soil. Promising treatments with various biological organisms have recently been reported for reducing disease severity and improving nutrient uptake and plant growth. These organisms include the root-nodulating symbiont *Rhizobium leguminosarum* (Frank) Frank bv. *phaseoli* (of D. C. Jordan in Bergey's Manual), and the vesicular-arbuscular mycorrhizal *Glomus mosseae* (Nicolson & Gerdemann) Gerdemann & Trappe. Soil fumigation with chemicals, such as chloropicrin or methyl bromide, also can manage the disease but is rarely cost effective.

Because soil conditions greatly affect Fusarium root rot, cultural practices can be employed to counteract the disease. Good soil fertility is important in maximizing bean yields, whether or not root rot is a problem. However, the form or quantity of nitrogen or other fertilizer elements has not been found to either stimulate or reduce the disease incidence in the field. Mounding soil around the plants can increase adventitious root formation and reduce the damaging effects of the disease on plant productivity. Wide spacing of plants within the row decreases spread between root systems, but plant populations that provide complete ground cover often give the highest seed yields. Beans grown in warm soil (20°C and higher) with near-optimal soil water potential usually suffer little from Fusarium root rot. Several green manure crops, when plowed into the soil, reduced disease severity from Fusarium root rot in subsequent kidney bean crops. The strongest antifungal activity was observed with red clover and alfalfa as precrops.

Minimizing soil compaction is probably the most effective means of root rot management. This can be achieved by loosening sublayers or wheel tracks with chisels before or at planting time, by not cultivating wet soil, and by reducing the pressure exerted by wheels on the soil surface. The addition of large amounts of organic matter to the soil by rotation of beans with crops such as small grains and alfalfa tends to counteract root

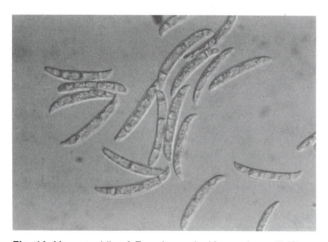

Fig. 14. Macroconidia of *Fusarium solani* f. sp. *phaseoli*. (Courtesy R. Hall)

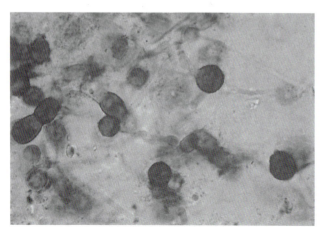

Fig. 15. Chlamydospores of *Fusarium solani* f. sp. *phaseoli*. (Courtesy R. Hall)

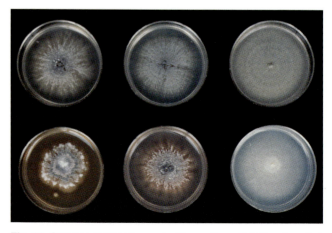

Fig. 16. Cultures of *Fusarium solani* f. sp. *phaseoli* (left and center) and *F. solani* (right) on potato dextrose agar. (Courtesy R. Hall)

rot by reducing compaction and increasing the water-holding capacity of the soil. Decaying alfalfa roots provide channels for deep penetration of the soil by bean roots. Sometimes, however, nondecomposed organic matter from almost any previous crop may be toxic to bean roots and increase root rot.

Bean genotypes differ in degree of sensitivity to Fusarium root rot. None are known to be highly resistant. Bush-type beans with less-vigorous root systems generally suffer more root rot than do indeterminate types with larger, more-vigorous root systems. Cultivars with resistance or tolerance to the disease or to predisposing factors have recently become available. Rapid improvement of resistance may now be enhanced with the discovery of resistance in the bean genome associated with certain host defense responses, including loci controlling pathogenesis-related proteins. Utilization of resistant or tolerant beans together with practicable management of soil nutrition, moisture, and compaction manage Fusarium root rot effectively.

Selected References

Buonassi, A. J., Copeman, R. J., Pepin, H. S., and Eaton, G. W. 1986. Effect of *Rhizobium* spp. on *Fusarium solani* f. sp. *phaseoli*. Can. J. Plant Pathol. 8:140-146.

Burke, D. W., and Miller, D. E. 1983. Control of Fusarium root rot with resistant beans and cultural management. Plant Dis. 67:1312-1317.

Dar, G. H., Zargar, M. Y., and Beigh, G. M. 1997. Biocontrol of Fusarium root rot in the common bean (*Phaseolus vulgaris* L.) by using symbiotic *Glomus mossea* and *Rhizobium leguminosarum*. Microb. Ecol. 34:74-80.

Kraft, J. M., Burke, D. W., and Haglund, W. A. 1981. *Fusarium* diseases of beans, peas, and lentils. Pages 142-156 in: *Fusarium* Diseases, Biology, and Taxonomy. P. E. Nelson, T. A. Tousson, and R. J. Cook, eds. Pennsylvania State University Press, University Park.

Okumura, M., Higashida, S., Yamagami, M., and Shimono, K. 1994. Effects of different preceding crops on Fusarium root rot of kidney bean. Jpn. J. Soil Sci. Plant Nutr. 65:274-281.

Roman-Aviles, B., Snapp, S. S., and Kelly, J. D. 2004. Assessing root traits associated with root rot resistance in common bean. Field Crops Res. 86:147-156.

Schneider, K. A., Grafton, K. F., and Kelly, J. D. 2001. QTL analysis of resistance to Fusarium root rot in bean. Crop Sci. 41:535-542.

Sippell, D. W., and Hall, R. 1982. Effects of pathogen species, inoculum concentration, temperature, and soil moisture on bean root rot and plant growth. Can. J. Plant Pathol. 4:1-7.

Snapp, S., Kirk, W., Roman-Aviles, B., and Kelly, J. 2003. Root traits play a role in integrated management of Fusarium root rot in snap bean. HortScience 38:187-191.

(Prepared by D. W. Burke and R. Hall;
Revised by R. M. Harveson and G. Yuen)

Fusarium Wilt (Yellows)

Fusarium wilt or yellows was originally discovered in dry beans in California in 1928. It has since been found elsewhere in the United States, South and Central America, Spain, and Africa. The disease is becoming more important in the midwestern United States and is considered to be important in Brazil. A similar disease of scarlet runner bean has been reported in England and the Netherlands.

Symptoms

Initial symptoms are slight yellowing and premature senescence of the lower leaves. The chlorotic symptoms progress up the plant until all leaves are bright yellow (Fig. 17), followed by wilting and discoloration (tan to brown) of foliage. If plants are infected when young, they remain stunted. The vascular tissues usually become reddish brown, often extending beyond the second node (Fig. 18).

Causal Organism

Fusarium yellows is caused by the fungus *Fusarium oxysporum* Schlechtend.:Fr. f. sp. *phaseoli* J. B. Kendrick & W. C. Snyder. At least seven pathogenic races are known. Maximum mycelial growth occurs on culture medium at 28°C. The fungus typically has hyaline, nonseptate chlamydospores (2–4 × 6–15 µm) and macroconidia that are elongate, have two or three septa (3–6 × 25–35 µm), and are slightly curved.

Disease Cycle and Epidemiology

The pathogen inhabits soil in the form of chlamydospores and may also infest seeds. Although symptoms occur only on *Phaseolus* spp., the pathogen can colonize the roots of other plants, particularly legumes, and produce chlamydospores without causing symptoms or disease. Infection of *Phaseolus* beans occurs through roots and hypocotyls, most commonly through wounds. Thereafter, the fungus grows throughout and plugs the vascular tissue, causing the plant to become chlorotic and drop its leaves or wilt.

The optimum temperature for disease development is 20°C. Extremes of soil moisture levels do not appear to be needed

Fig. 17. Fusarium yellows, caused by *Fusarium oxysporum* f. sp. *phaseoli*. (Courtesy H. F. Schwartz)

Fig. 18. Reddish brown discoloration of vascular tissue of a plant affected by Fusarium yellows, caused by *Fusarium oxysporum* f. sp. *phaseoli*. (Courtesy H. F. Schwartz)

for the disease to occur but can influence disease severity. Soil compaction and poor drainage also appear to aggravate disease severity.

Management

Little information is available on management of the disease. Tolerant or resistant cultivars are recommended where available. Race-specific resistance conferred by single to multiple genes from different races of beans has been incorporated with conventional breeding and molecular techniques into various bean cultivars. This resistance is effective against many of the seven or more pathogenic races that occur worldwide. Crop rotation may help in reducing soil inoculum levels if reservoir hosts are avoided. Chemical seed treatment and reduction of soil compaction also may be useful.

Selected References

Abawi, G. S. 1989. Root rots. Pages 105-157 in: Bean Production Problems in the Tropics, 2nd ed. H. F. Schwartz and M. A. Pastor-Corrales, eds. Centro Internacional de Agricultura Tropical (CIAT), Cali, Colombia.

Alves-Santos, F. M., Benito, E. P., Eslava, A. P., and Diaz-Minguez, J. M. 1999. Genetic diversity of *Fusarium oxysporum* strains from common bean fields in Spain. Appl. Environ. Microbiol. 65:3335-3340.

Cramer, R. A., Byrne, P. F., Brick, M. A., Panella, L., Wickliffe, E., and Schwartz, H. F. 2003. Characterization of *Fusarium oxysporum* isolates from common bean and sugar beet using pathogenicity assays and random-amplified polymorphic DNA markers. J. Phytopathol. 151:352-360.

Dhingra, O. D., and Coelho Netto, R. A. 2001. Reservoir and non-reservoir hosts of bean-wilt pathogen, *Fusarium oxysporum* f. sp. *phaseoli*. J. Phytopathol. 149:463-467.

Fall, A. L., Byrne, P. F., Jung, G., Coyne, D. P., Brick, M. A., and Schwartz, H. F. 2001. Detection and mapping of a major locus for Fusarium wilt resistance in common bean. Crop Sci. 41:1494-1498.

Ribeiro, R. de L. D., and Hagedorn, D. J. 1979. Screening for resistance to and pathogenic specialization of *Fusarium oxysporum* f. sp. *phaseoli*, the causal agent of bean yellows. Phytopathology 69:272-276.

Salgado, M. O., Schwartz, H. F., and Brick, M. A. 1995. Inheritance of resistance to a Colorado race of *Fusarium oxysporum* f. sp. *phaseoli* in common beans. Plant Dis. 79:279-281.

(Prepared by D. J. Hagedorn;
Revised by H. F. Schwartz and G. Yuen)

Phymatotrichum Root Rot

Phymatotrichum root rot, also commonly known as Texas root rot, cotton root rot, and Ozonium root rot, is found primarily in alkaline, calcareous soils of the southwestern United States and northern Mexico. The fungus has been reported in India, Russia, and the United States (Hawaii), but disease losses have not been reported in those areas. The pathogen infects more than 2,000 species of dicotyledonous plants, and all beans are very susceptible.

Symptoms

The first symptom of the disease is slight yellowing or bronzing of the leaves, followed by sudden wilting when plants begin to flower. Cortical tissue is usually killed, sloughs easily, and is covered by a visible network of hyphal strands. Plants usually die within a few days after wilting, often in a circular pattern, as the fungus grows radially from dying plants. After rains, spore mats of the fungus may occur on the soil surface around the stems of dead plants.

Causal Organism

Phymatotrichum omnivorum Duggar (syn. *Phymatotrichopsis omnivora* (Duggar) Hennebert), also referred to as *Ozonium omnivorum* Shear in older literature, produces rhizomorphlike strands in soil that are composed of a large central hypha entwined by many smaller hyphae. Hyphae have characteristic branches at right angles to the strands, often referred to as cruciform branching. Sclerotia are formed in coarse portions of strands away from a food base. A conidial state (similar to *Botrytis* spp.) is formed on spore mats after frequent rains and cloudy days. Conidia are globose, single-celled, and 4.5–5.0 µm in diameter; they usually do not germinate in culture. The teleomorph has been reported as *Hydnum omnivorum* Shear and as *Sistotrema brinkmannii* (Bres.) J. Eriksson. In another case, a basidial stage produced in vitro was named *Trechispora brinkmannii* (Bres.) D. P. Rogers & H. Jackson.

Disease Cycle and Epidemiology

As plants die, sclerotia form on rhizomorphlike strands in the roots. The sclerotia persist deep (45–75 cm) in the soil or in the living roots of susceptible hosts. Strands of the fungus grow outward from sclerotia until they contact a descending root. They entwine the root and grow toward the soil surface. When the strands grow around the upper root system, they proliferate and form a cottony growth of hyphae around the plant. The fungus invades the roots through wounds or directly through the cortex. It can invade the inner root tissues, blocking the flow of water and photosynthates through the vascular system.

Environmental fluctuations cause disease severity to vary from year to year. The disease is rarely found in acidic soils (pH <6.0) or in soils with less than 1% calcium carbonate. The pathogen may be introduced on the roots of various plants, including transplanted trees and ornamentals. The fungus is favored by warm (25–35°C), moist soils and thrives in heavier soils with irrigation. It persists where the temperature does not fall below –23°C, where the mean annual temperature exceeds 15°C, and where a frost-free period of 200 days annually occurs. It is thought that high CO_2 levels and the formation of bicarbonate in poorly drained soils promote pathogen growth and suppress competitive microorganisms. If the fungus is introduced into sandy, acidic soils, it grows and kills plants, but it does not produce sclerotia and persist.

Management

It is very difficult to manage Phymatotrichum root rot because the fungus behaves erratically and disease severity varies from year to year. No resistance to the pathogen is known in beans. Monocotyledons are less susceptible to the pathogen than are dicotyledons, and a 4-year rotation with corn, sorghum, or grasses is recommended. Weeds should be managed. Clean fallow for 1 year delays infection and reduces disease severity. Deep-chisel or deep-plow tillage and organic soil amendments, such as green manure crops, animal manure, or composts, may reduce the disease severity. Planting rows of monocotyledons or digging barriers adjacent to infested areas keeps the fungus from growing into surrounding soil. Soil fumigation or deep injection of ammonia to kill the fungus may be justified in a high-value crop.

Selected References

Baniecki, J. F., and Bloss, H. E. 1969. The basidial stage of *Phymatotrichum omnivorum*. Mycologia 61:1054-1059.

Lyda, S. A. 1978. Ecology of *Phymatotrichum omnivorum*. Annu. Rev. Phytopathol. 16:193-209.

Streets, R. B., and Bloss, H. E. 1973. Phymatotrichum Root Rot. Monogr. 8. American Phytopathological Society, St. Paul, MN.

(Prepared by D. R. Sumner; Revised by R. M. Harveson)

Pythium Diseases

Pythium infections in beans cause seed rot (Fig. 19), pre-emergence and postemergence damping-off (Fig. 20), stem rot (Fig. 21), root rot (Figs. 19 and 22), blight (Fig. 23), pod rot, and stunting (Fig. 24). Yield losses can range up to 100%, but rarely exceed 20%. Since the pathogen is widely distributed in soils, these diseases can occur wherever beans are grown.

Symptoms

Poor stands attributable to *Pythium* infection can be the result of seed rot (Fig. 19) or preemergence or postemergence damping-off of seedlings (Fig. 20). The affected tissue becomes mushy and discolored. Infected seedlings that do emerge may wilt and die within early (Fig. 23) or late (Figs. 25 and 26) stages of growth. These plants commonly show a lesion, initially water-soaked but becoming necrotic, extending from the roots up the hypocotyl (Fig. 21) and sometimes reaching the growing point. Infections that develop more slowly, although not fatal, may severely stunt the plant (Fig. 24). In these cases, the root system is necrotic and reduced in size. A necrotic lesion often extends up the hypocotyl for several centimeters above the soil level (Fig. 5). Affected plants usually produce adventitious roots from the hypocotyl. Mild infections, in which the feeder roots are attacked (Fig. 22) but the shoots are symptomless, are common and undoubtedly reduce plant productivity.

Fig. 19. Seed decay (left) and root rot (right) caused by *Pythium paroecandrum*. (Courtesy R. Hall)

Fig. 20. Postemergence damping-off caused by *Pythium paroecandrum*. (Courtesy R. Hall)

Fig. 21. Lesions on hypocotyls caused by a *Pythium* sp. (Courtesy H. F. Schwartz)

Fig. 22. Rotting of feeder roots caused by *Pythium paroecandrum*. (Courtesy R. Hall)

Fig. 23. Blight of a bean seedling caused by a *Pythium* sp. (Courtesy H. F. Schwartz)

Under irrigation or during cool and prolonged moist conditions, pods in contact with the soil may become infected, exhibiting water-soaking and fluffy white fungal growth. This disease may be mistaken for the early stages of white mold, caused by *Sclerotinia sclerotiorum* (Lib.) de Bary. However, with *Pythium* spp. no sclerotia are produced, while *S. sclerotiorum* produces sclerotia.

Causal Organisms

The *Pythium* spp. of major importance as bean pathogens can be placed into two groups according to their morphology and response to temperature. The first group contains species, such as *P. ultimum* Trow, *P. irregulare* Buisman, and *P. paroecandrum* Drechs., that produce spherical oospores and spherical sporangia (Fig. 27), and the second group are those that resemble *P. ultimum* but lack oospores. Members of the latter group appear to be the major pathogenic species in some northern areas of North America (e.g., New York, Ontario, and Wisconsin) and are most active at temperatures below 25°C. In other areas of the United States (e.g., Maryland and Georgia), *P. myriotylum* Drechs. and *P. aphanidermatum* (Edson) Fitzp. predominate. These species have lobate sporangia and are most active at 20–35°C. Other *Pythium* spp. that have been reported on beans but appear to be of minor importance are *P. acanthicum* Drechs., *P. anandrum* Drechs., *P. aristosporum* Vanterpool, *P. dissotocum* Drechs., *P. helicoides* Drechs., *P. mamillatum* Meurs, *P. oligandrum* Drechs., *P. pulchrum* Minden, *P. rostratum* E. J. Butler, *P. spinosum* Sawada, and *P. vexans* de Bary. Oogonia and antheridia join to produce spherical oospores, which can range from 10 to 40 μm in diameter, depending on the species.

Pythium spp. can generally be isolated on water agar and on more-complex selective media. Identification is facilitated by allowing the fungus to colonize pieces of boiled grass leaves placed on the culture medium and then floating the grass in water to allow production of sporangia.

Disease Cycle and Epidemiology

Beans can be attacked by several *Pythium* spp., all of which are similar with respect to disease cycle. The fungus survives in soil as thick-walled oospores and, in some species (e.g., *P. ultimum*), as sporangia. These spores remain quiescent until stimulated to germinate by an exogenous source of nutrients, such as seed or root exudates. After infecting the plant, secondary infections may arise from zoospores released from sporangia. Zoospores can swim a limited distance in water films in soil, thus concentrating inoculum at an infection court or permitting the fungus to reach surface water, in which it may move a greater distance. Oospores formed in infected tissue are released to the soil when the tissue is further decayed. Oospores can be moved long distances in infested soil that is carried by wind or water or on contaminated equipment.

Pythium spp. are ubiquitous in soil because of their broad host ranges, their ability to colonize saprophytically fresh organic

Fig. 24. Stunting of bean plants caused by a *Pythium* sp. Healthy plant is on the left. (Courtesy H. F. Schwartz, from the files of G. S. Abawi)

Fig. 25. Pythium blight caused by *Pythium ultimum.* (Courtesy D. J. Hagedorn)

Fig. 26. Pythium blight caused by *Pythium* spp. (Courtesy R. L. Forster)

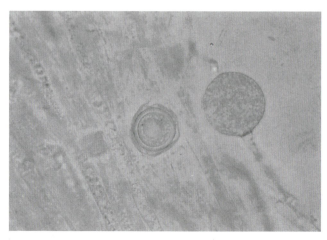

Fig. 27. Sporangium (right) and oogonium containing an oospore (left) of *Pythium paroecandrum* in a bean root. (Courtesy R. Hall)

matter in soil, and their persistent oospores. Their populations can increase on a number of crops, and severity of *Pythium* diseases in beans can increase with increasing inoculum levels.

Soil moisture and temperature have a profound influence on the incidence and severity of these diseases. Disease severity increases with increasing soil moisture levels. Wet soil conditions favor damping-off and the extension of hypocotyl lesions up to the growing point of the young plant. The two groups of *Pythium* spp. described above respond differently to temperature. Disease incited by *P. ultimum* and similar species is more common and more severe at temperatures below 20°C. Pythium blight incited by *P. myriotylum*, on the other hand, does little damage below 20°C and becomes severe at 25–35°C. *Pythium* diseases can occur throughout the season, whenever environmental conditions favor them. Even pods on plants approaching maturity can become infected; severity of the disease then depends largely on moisture and temperature conditions. Infections on older plants are restricted unless moisture is abundant.

Pythium spp. commonly occur as a member of a disease complex that may include *Fusarium solani* (Mart.) Sacc. f. sp. *phaseoli* (Burkholder) W. C. Snyder & H. N. Hans. and species of genera *Rhizoctonia*, *Thielaviopsis*, and *Aphanomyces*. Synergistic interactions that increase disease severity can occur between *P. ultimum* and *F. solani*, as well as between *P. ultimum* and *Aphanomyces* spp. (Figs. 6 and 7).

Management

Chemical seed treatment is currently the major management method for seed decay and damping-off caused by *Pythium* spp. Several nonsystemic seed dressings can provide adequate management of these diseases under mild disease pressure. Systemic fungicides, such as mefenoxam, also provide some protection against the root rot and blight stages of the disease. However, *P. myriotylum* and *P. aphanidermatum* are less sensitive to mefenoxam than are the *P. ultimum*-type fungi. Herbicide interaction with *Pythium* spp. is variable, with some having a suppressive effect and glyphosate having a predisposing effect.

Many organisms have been demonstrated to provide some antagonism to or management of *Pythium* spp. Various biological control organisms are available commercially with antifungal activity, including formulations with *Bacillus* spp., *Burkholderia cepacia* (Palleroni & Holmes) Yabuuchi et al., *Streptomyces griseoviridis* Anderson et al., *Gliocladium virens* J. H. Miller, J. E. Giddens, & A. A. Foster, and *Trichoderma harzianum* Rifai. Biological control organisms, such as *Bacillus subtilis* (Ehrenberg) Cohn, may perform similar to chemical fungicides but frequently are less consistent in their effectiveness because of environmental variables. Perhaps the most efficient method of utilizing seed treatments for managing *Pythium* diseases may be the integration of both techniques.

Cultural practices that alleviate wet, poorly drained soils can also influence the incidence and severity of Pythium root rot. Several reports indicate less disease following moldboard plowing than after minimum tillage. In one field study, a combination of plowing and metalaxyl seed treatment gave lower disease severity than either management measure alone. Avoidance of heavily infested fields by testing soil for root rot potential, addition of various composting amendments, and crop rotation are measures that can reduce losses.

Genetic resistance to *Pythium* diseases in snap beans has not been very effective and has been difficult to increase. Beans with colored seeds show good levels of resistance to seed rot and damping-off, but this resistance has been very difficult to transfer to agronomically acceptable, white-seeded bean cultivars. This problem arises from the apparently polygenic nature of the resistance and linkage with undesirable traits, such as late maturity, low yield potential, and poor pod quality. Furthermore, resistance to damping-off is not necessarily correlated with resistance to the root rot phase of the disease. High levels of resistance to *Pythium* diseases have been found in some white-seeded snap bean lines.

Selected References

Abawi, G. S. 1989. Root rots. Pages 105-157 in: Bean Production Problems in the Tropics, 2nd ed. H. F. Schwartz and M. A. Pastor-Corrales, eds. Centro Internacional de Agricultura Tropical (CIAT), Cali, Colombia.

Descalzo, R. C., Punja, Z. K., Levesque, C. A., and Rahe, J. E. 1996. Identification and role of *Pythium* species as glyphosate synergists on bean (*Phaseolus vulgaris*) grown in different soils. Mycol. Res. 100:971-978.

Lewis, J. A., Lumsden, R. D., Papavizas, G. C., and Kantzes, J. G. 1983. Integrated control of snap bean diseases caused by *Pythium* spp. and *Rhizoctonia solani*. Plant Dis. 67:1241-1244.

Lumsden, R. D., Ayers, W. A., Adams, P. B., Dow, R. L., Lewis, J. A., Papavizas, G. C., and Kantzes, J. G. 1976. Ecology and epidemiology of *Pythium* species in field soil. Phytopathology 66:1203-1209.

Pfender, W. F., and Hagedorn, D. J. 1982. Comparative virulence of *Aphanomyces euteiches* f. sp. *phaseoli* and *Pythium ultimum* on *Phaseolus vulgaris* at naturally occurring inoculum levels. Phytopathology 72:1200-1204.

Pieczarka, D. J., and Abawi, G. S. 1978. Effect of interaction between *Fusarium*, *Pythium*, and *Rhizoctonia* on severity of bean root rot. Phytopathology 68:403-408.

Schuler, C., Biala, J., Bruns, C., Gottschall, R., Ahlers, S., and Vogtmann, H. 1989. Suppression of root rot of peas, beans, and beet roots caused by *Pythium ultimum* and *Rhizoctonia solani* through the amendment of growing media with composted organic household waste. J. Phytopathol. 127:227-238.

Sumner, D. R., Smittle, D. A., Threadgill, E. D., Johnson, A. W., and Chalfant, R. B. 1986. Interactions of tillage and soil fertility with root diseases in snap bean and lima bean in irrigated multiple-cropping systems. Plant Dis. 70:730-735.

(Prepared by W. F. Pfender and D. J. Hagedorn; Revised by R. M. Harveson)

Rhizoctonia Root Rot

Rhizoctonia root rot of beans is common throughout the world and is one of the most economically important root and hypocotyl diseases of beans in large and small plantings. Losses of more than 10% in conventional tillage and 20–30% in minimal or no-till systems have occurred in the United States; and in Brazil, up to 60% yield reductions have been reported in conjunction with Fusarium root rot. Rhizoctonia root rot also is an important disease on a large number of other crop plants.

Symptoms

Small, elongate, sunken, reddish brown lesions on hypocotyls and roots are typical symptoms of early disease development (Fig. 28). As these lesions increase in size and become more sunken, they become cankerous, and the red color may predominate until the cankers are old. Hypocotyls are often girdled by the coalescence of several cankers, resulting in preemergence or postemergence damping-off. Severe infections cause plant stunting and premature death. Small brown-black sclerotia may form on the surface (or just beneath the surface) of older cankers that may be rough and dry. Occasionally, the fungus enters and destroys the pith (Fig. 29).

Causal Organism

This disease is caused by the soilborne fungus *Rhizoctonia solani* Kühn, particularly anastomosis groups AG-4 and AG-2-2. *Thanatephorus cucumeris* (A. B. Frank) Donk is the teleomorph. Web blight, discussed in this compendium, is also caused by *R. solani* and *T. cucumeris*. This organism exists primarily as sterile mycelium that is colorless when young and becomes yellowish and then brown when older. Its cells are long and multinucleate. Branch hyphae arise at right angles, are constricted at their base, and contain a cross-wall near the branch point. Tufts of broad, short, oval to triangular cells may form and act directly as long-term survival structures or develop into dark brown sclerotia.

The basidiomycete teleomorph is less common. It forms blackish, barrel-shaped basidia under conditions of high humidity on a thin, membranous layer of mycelium on the soil or on plant parts near the soil level. Four sterigmata are typical, each bearing an oval basidiospore.

Disease Cycle and Epidemiology

R. solani is a collective species, and isolates vary considerably in host preference and response to environmental conditions. Generally, the pathogen survives between crop seasons as sclerotia or mycelium in soil, particularly in the upper 15–20 cm; in or on infested plant debris; or on perennial plants. It may be borne on or within bean seeds. It is usually disseminated in infested soil or plant debris by wind, rain, irrigation water, and farm implements. When soils become infested, they remain so indefinitely.

The pathogen can infect the host in two ways. Infection pegs produced from infection cushions can penetrate the intact cuticle and epidermis, and individual hyphae can penetrate the plant through natural openings and wounds. Once infection has taken place, the fungus ramifies rapidly through adjacent cells and tissues. Pectolytic and cellulolytic enzymes produced by the pathogen are important in its pathogenicity.

The disease is most severe at 15–18°C. At 21°C, numbers of cankers are substantially reduced, perhaps because plants emerge rapidly and escape infection. At temperatures below the optimum (down to 9°C), lesions are fewer and smaller, possibly because of direct effects of temperature on the fungus. Soil moisture conditions have little effect on disease severity, but soil compaction increases root rot incidence and severity.

Plant age plays an important role in the epidemiology of Rhizoctonia root rot of beans. Seedlings and young plants are highly susceptible to infection, whereas new infections in plants older than 4 weeks are rare or have minimal effect.

Management

Resistance to Rhizoctonia root rot is available in bean breeding lines and moderate resistance is found in some commercial cultivars. A number of fungicides can manage Rhizoctonia root rot on young bean plants if applied as a seed treatment or soil drench. The incidence of Rhizoctonia root rot may be reduced by shallow seeding, rotation of beans with nonhost crops, and planting in warm soil. Working the soil to reduce compaction also can reduce disease severity. Sugar beet and potato may increase fungal inoculum and should be avoided in rotations where Rhizoctonia root rot has been a problem. Dumping beet dirt where beans are to be planted should be avoided.

Selected References

Abawi, G. S. 1989. Root rots. Pages 105-157 in: Bean Production Problems in the Tropics, 2nd ed. H. F. Schwartz and M. A. Pastor-Corrales, eds. Centro Internacional de Agricultura Tropical (CIAT), Cali, Colombia.

Engelkes, C. A., and Windels, C. E. 1996. Susceptibility of sugar beet and beans to *Rhizoctonia solani* AG-2-2 IIIB and AG-2-2 IV. Plant Dis. 80:1413-1417.

Kramer, N., Hagedorn, D. J., and Rand, R. E. 1975. *Rhizoctonia solani* seed-borne on *Phaseolus vulgaris*. Proc. Am. Phytopathol. Soc. 2:42.

Muyolo, N. G., Lipps, P. E., and Schmitthenner, A. F. 1993. Reactions of dry bean, lima bean, and soybean cultivars to Rhizoctonia root and hypocotyl rot and web blight. Plant Dis. 77:234-238.

Prasad, K., and Weigle, J. L. 1969. Resistance to *Rhizoctonia solani* in *Phaseolus vulgaris* (snap bean). Plant Dis. Rep. 53:350-352.

Sneh, B., Burpee, L., and Ogoshi, A. 1991. Identification of *Rhizoctonia* species. American Phytopathological Society, St. Paul, MN.

Tu, J. C., and Tan, C. S. 1991. Effect of soil compaction on growth, yield and root rots of white beans in clay loam and sandy loam soil. Soil Biol. Biochem. 23:233-238.

(Prepared by D. J. Hagedorn; Revised by L. E. Hanson)

Fig. 28. Sunken, reddish brown cankers on bean hypocotyls caused by *Rhizoctonia solani*. (Courtesy J. Springer, from the files of L. Hanson)

Fig. 29. Reddish brown rotting of the pith of bean stems caused by *Rhizoctonia solani*. (Courtesy H. F. Schwartz, from the files of S. E. Beebe)

Southern Blight

Southern blight, also called southern stem rot, southern wilt, or crown rot, is an important disease that is widespread in subtropical and tropical areas of the United States and throughout the world where high temperatures and wet weather prevail during the growing season. Southern blight occasionally occurs in temperate zones, but losses usually are minor. The fungus causing southern blight infects more than 500 species of monocotyledons and dicotyledons but is especially severe on legumes, solanaceous crops, cucurbits, and other vegetables grown in rotation with beans.

Symptoms

The earliest symptoms of southern blight are a slight yellowing of the lower leaves and water-soaking and slight darkening of the hypocotyl just below the soil line, followed by a yellowing of the upper leaves and leaf drop. The fungus destroys the cortex and grows downward in the stem and roots (Fig. 30). It occasionally invades the vascular tissues and grows systemically upward into the lower branches, causing a dark discoloration of the tissue. The pathogen girdles the stem at the soil line, causing wilting and death (Fig. 31). A weft of coarse mycelium may form on the stem and spread into surrounding soil, organic matter, and pods in contact with the soil. Characteristic spherical brown sclerotia form on the mycelium and the base of the plant (Fig. 30).

Causal Organism

The anamorphic state of the causal fungus is *Sclerotium rolfsii* Sacc. Hyphae of *S. rolfsii* are coarse, have clamp connections, and consist of cells that are $2–9 \times 150–250$ µm. Sclerotia are 0.5–1.5 mm in diameter. The rarely seen teleomorph, *Athelia rolfsii* (Curzi) Tu & Kimbrough, produces an exposed hymenium with clavate basidia and hyaline, pyriform basidiospores that are $1.0–1.7 \times 6–12$ µm.

Disease Cycle and Epidemiology

Since basidia are rare, the fungus is disseminated by mycelium in infested organic matter and by sclerotia in soil. Infection may occur by hyphae directly penetrating host tissues at any time after seedling emergence. Infection usually takes place at the soil surface but may occur belowground in light, sandy soils. The fungus ramifies through the cortex and into the vascular tissues. Oxalic acid, pectic enzymes, and cellulolytic enzymes are produced, and tissues are destroyed, usually causing plant death. The fungus may spread more than 1 m through the soil and from plant to plant within a row. Sclerotia produced on organic matter and dying plants serve as inoculum for the next crop.

Fig. 30. Lower stem and root lesions of southern blight (right) and sclerotia produced by *Sclerotium rolfsii* (left). (Courtesy H. F. Schwartz, from the files of M. A. Pastor-Corrales)

Fig. 31. Lesions on lower stems of Eagle snap beans at 16–24°C caused by *Sclerotium rolfsii* (left) and by a sterile basidiomycete (right). (Courtesy D. J. Hagedorn, from the files of D. R. Sumner)

The optimum temperature for growth of *S. rolfsii* is 30°C. Growth decreases markedly below 15°C and above 37°C. Low temperatures prevent the disease from being a serious threat to bean production in temperate zones. Moist, warm weather and abundant organic matter on the soil surface are necessary for the pathogen to cause severe disease. The fungus requires a readily available food source to have sufficient energy for parasitism. Cultivators and other farm equipment spread sclerotia and dead plants within a field. The pathogen may also be distributed in diseased transplants and in irrigation water, runoff, and soil from infested fields. Numerous susceptible weeds are sources of inoculum. If infested bean fields are grazed by livestock, sclerotia can pass through the digestive tract of cattle and sheep without loss of viability. Sclerotia rapidly lose viability in moist soil but survive for months in organic matter underwater or saturated with water.

Management

Turning the soil to a depth of 20–25 cm with a moldboard plow buries infested debris and sclerotia and limits contact of the fungus with bean hypocotyls and stems. Cultivation that moves soil into contact with plants should be avoided, and susceptible weeds should be removed by hand or managed with herbicides. Crop rotations do not eliminate the pathogen, but rotations with corn, sorghum, small grains, or grasses reduce disease severity. Registered fungicides can be applied in the furrow or over the row at planting. In subtropical regions, early planting when the soil is cool lessens the likelihood of infection. Cultivars resistant to *S. rolfsii* are available. In warm, dry climates, soil solarization may kill *S. rolfsii* in the topsoil. In the tropics, mulches have been effective in reducing wilt and increasing yield. A large number of different bacterial and fungal organisms has demonstrated some promise as antagonists for reducing the effects of diseases caused by *S. rolfsii*. These microorganisms have been tested as both seed and soil treatments, and species of the genera *Gliocladium* and *Trichoderma* are notable as the most commonly evaluated candidates as biofungicides for management of *S. rolfsii*.

Selected References

Abawi, G. S. 1989. Root rots. Pages 105-157 in: Bean Production Problems in the Tropics, 2nd ed. H. F. Schwartz and M. A. Pastor-Corrales, eds. Centro Internacional de Agricultura Tropical (CIAT), Cali, Colombia.

Beebe, S. E., Bliss, F. A., and Schwartz, H. F. 1981. Root rot resistance in common bean germ plasm of Latin American origin. Plant Dis. 65:485-489.

Lewis, J. A., and Fravel, D. R. 1996. Influence of Pyrax/biomass of biocontrol fungi on snap bean damping-off caused by *Sclerotium rolfsii* in the field and on germination of sclerotia. Plant Dis. 80:655-659.

Lewis, J. A., Papavizas, G. C., and Hollenbeck, M. D. 1993. Biological control of damping-off of snapbeans caused by *Sclerotium rolfsii* in the greenhouse and field with formulations of *Gliocladium virens*. Biol. Control 3:109-115.

Roberti, R., Flori, P., and Pisi, A. 1996. Biological control of soilborne *Sclerotium rolfsii* infection by treatment of bean seeds with species of *Trichoderma*. Petria 6:105-116.

(Prepared by D. R. Sumner; Revised by R. M. Harveson)

Stem Rot

A stem rot caused by an unidentified basidiomycete was found in Florida in 1968 and has since been described on snap beans in Georgia, pigeon pea in Puerto Rico, chickpea in India, and most recently, on dry edible beans (great northern) in Nebraska. This type of fungus causes root disease in nature on many other hosts, including corn sorghum, cowpea, peanut, soybean, and squash. Also, the fungus is pathogenic on rye, onion, spinach, cucumber, and watermelon. There are unsubstantiated reports of a similar fungus in other countries in the subtropics and tropics. The disease is probably widespread in warm climates but is of minor importance on beans.

Symptoms

The first symptoms of disease after emergence are wilting, followed by plant death within a few days. On less severely infected plants, small lesions may be scattered on the hypocotyls and fibrous roots. The disease in Georgia is similar to southern blight, except that stem lesions are usually dry and firm (Fig. 31). Hypocotyl and stem lesions may be superficial, tan lesions or sunken, gray to black cankers. Reddish dark brown cortical lesions reminiscent of Fusarium root rot characterized symptoms observed from infected dry beans in Nebraska. White mycelial strands may grow over the lesions, and when wilted plants are pulled, a collar of soil and mycelium adheres to the plant at the soil line. No sclerotia or wefts of mycelium occur on the stem and surrounding soil, in contrast to symptoms of southern blight, caused by *Sclerotium rolfsii* Sacc. If infection occurs during germination, preemergence damping-off may occur.

Causal Organism

The pathogen is a sterile, white basidiomycete (Fig. 32) with prominent clamp connections, binucleate mycelium, wide-angle branching of the hyphae, and dolipore septa. In culture, submerged hyphae have predominantly monilioid cells, in contrast to the slender, mostly straight cells of *S. rolfsii*. Electrophoretic profiles of isozymes of aminopeptidase from the fungus and from *S. rolfsii* are distinctly different. One culture of a sterile, white basidiomycete isolated from bermudagrass formed sporophores in flasks of living sweet corn and snap bean plants and were identified as species of the genus *Marasmius*.

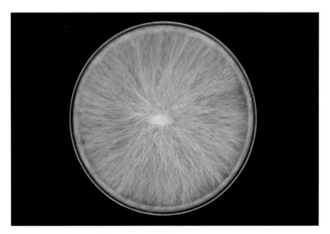

Fig. 32. Sterile, white basidiomycete, cause of stem rot. (Courtesy D. J. Hagedorn, from the files of D. R. Sumner)

Disease Cycle and Epidemiology

The fungus is most frequent in snap beans and cowpea in irrigated, multiple-cropping systems following corn. It can survive at least 1 year in fallow soil and can cause disease at temperatures of 15–36°C. The low end of the temperature range is considerably below the optimum for *S. rolfsii*. However, the fungus apparently is favored by high temperatures in fields and is more commonly associated with late-planted crops of snap beans. The fungus probably survives in the soil in colonized roots, crowns, and stems of corn; other susceptible crops and weeds; and peanut seeds and shells. The fungus invades roots and crowns of susceptible crops and weeds early in the growing season. When plant residues from those crops are shallow-tilled into the soils, infested debris is spread and more organic matter is colonized, thereby increasing inoculum levels in the soil for subsequent crops of beans.

Management

The disease is most likely of minor importance, and management measures have not been developed. In Georgia, the disease was less severe in snap beans following corn with mold-board plowing (20–25 cm) than with shallow operations, such as harrowing. The fungus has also been isolated from potato and sugar beet in Nebraska, which are common crops (along with corn) used in rotation with dry beans. These observations, in concert with prior work in Georgia, illustrate the potential problems involving pathogen survival or increase in quantity in soils for all subsequent susceptible crops. Snap beans, lima bean, pole bean, cowpea, chickpea, and dry edible beans are susceptible, but nothing is known about variation in susceptibility among cultivars. Management of the pathogen with fungicides has not been investigated. Soil solarization would probably eliminate the pathogen in the topsoil.

Selected References

Baird, R. E., Wilson, J. P., and Sumner, D. R. 1992. Identity and pathogenicity of two *Marasmius* species from the sterile white basidiomycete complex. Plant Dis. 76:244-247.

Bell, D. K., and Sumner, D. R. 1984. Ecology of a sterile white pathogenic basidiomycete in corn, peanut, soybean, and snap bean field microplots. Plant Dis. 68:18-22.

Harveson, R. M. 2002. A new soilborne disease of dry, edible beans in western Nebraska. Plant Dis. 86:1051.

Haware, M. P., and Nene, Y. 1978. A root rot of chickpea seedlings caused by a sterile fungus. Legume Res. 1:65-68.

Howard, C. M., Conway, K. E., and Albregts, E. E. 1977. A stem rot of bean seedlings caused by a sterile fungus in Florida. Phytopathology 67:430-433.

Sumner, D. R., Bell, D. K., and Huber, D. M. 1979. Pathology, host range, and ecology of a sterile basidiomycete causing root disease on corn. Plant Dis. Rep. 63:981-985.

(Prepared by D. R. Sumner; Revised by R. M. Harveson)

Fungal Diseases of Aerial Parts

Alternaria Leaf and Pod Spot

Alternaria leaf and pod spot (black pod) is widespread; it is reported from Brazil, Canada, Costa Rica, Colombia, Chile, East Africa, England, Mexico, the United States, and Venezuela. The disease can affect all aerial parts of the plant as it approaches maturity. Senescent leaves and mature pods are highly susceptible. Economic losses result primarily from reduced prices for discolored seeds and pods. Losses of 12% have been reported in snap beans.

Symptoms

Small flecks or tiny water-soaked spots occur on infected green leaves and pods. Lesions on leaves appear as circular to irregular spots and flecks with a pale brown center and a dark brown margin surrounded by a chlorotic halo (Fig. 33). The lesions may develop concentric rings, and diseased tissue may become dry and brittle and fall out, leaving a shot hole in the leaf. As plants approach maturity, growth of the fungus on the surface gives infected leaves, pods, and stems a dark, moldy appearance. Diseased plants turn from brown to black within 1 week of maturation during periods of rain or high humidity. Dry seeds from diseased pods often bear a gray, moldy growth and show a gray to dark gray discoloration. Occasionally, the discolored seeds are streaked with brown. They are more likely to have wrinkled seed coats (Fig. 34), and they have a slightly lower germinability than do normal beans. Seed discoloration is a direct result of discoloration of maturing pods (Fig. 35).

Causal Organisms

The disease is caused by various species of the genus *Alternaria*, including *A. alternata* (Fr.:Fr.) Keissl. (syn. *A. tenuis* Nees), *A. brassicae* (Berk.) Sacc. f. *phaseoli* C. Brun, and *A. brassicicola* (Schwein.) Wiltshire. *A. alternata* has dark, simple, short, septate conidiophores that bear a simple or branched chain of conidia. Conidia (10–20 × 20–40 µm) are dark, typically have one to three transverse and longitudinal septa, are

Fig. 33. Alternaria leaf spot, caused by *Alternaria alternata*. (Courtesy S. K. Mohan)

Fig. 34. Seed discoloration caused by *Alternaria alternata* (left). Healthy seeds are on the right. (Courtesy J. C. Tu)

slightly beaked, and are elliptical or obovoid in shape. *A. alternata* is a leaf-inhabiting fungus common in bean fields. It is a weak parasite that colonizes the cavities of stomata but causes little further infection during the active growing phase of plants. The fungus thrives on senescent tissues, sporulates readily, and overwinters in infested plant debris. The discolored pods carry numerous spores of the causal fungi and contain a water-soluble, blackish brown pigment that smudges the dry seeds.

Disease Cycle and Epidemiology

A. alternata survives between crops as conidia in infested crop and weed debris and grows as an epiphyte or as a pathogen on beans and other plants during the following year. The fungus has been isolated from bean plants at all stages of growth and also from most weeds associated with beans. The weeds are typically symptomless hosts. Population densities of *A. alternata* on the leaves of weeds and beans increase as the growing season advances. The fungus thrives on mature and senescent plant tissue late in the season, especially under damp or wet conditions. Bean plants often turn from brown to dark gray when harvesting is delayed by rain.

Leaf exudates have an important role in the growth and sporulation of epiphytic fungi. The population of *A. alternata* has been correlated positively with natural senescence of leaf tissues. Sugars and ninhydrin-positive substances in the leaf wash increase with plant age and cultivar susceptibility to *A. alternata*. Black pod disease and seed coat discoloration are more common in early-maturing cultivars. Cultivar and plant age may affect susceptibility to Alternaria leaf and pod spot by determining the quantity and quality of plant secretions and, thereby, the population density of *A. alternata*.

In a random sample of dry bean seeds, *A. alternata* infested 88–94% and 15–36% of the seeds from discolored and nondiscolored seed lots, respectively. In an extensive survey of bean seeds in Ontario, Canada, 296 of 493 samples were discolored.

Management

Management of the disease in snap bean pods involves the use of wider row spacing, fungicides, resistant cultivars, and crop rotation. Management in dry beans also depends on these practices but relies mostly on regulating the moisture content of seeds. Seeds should be harvested as soon as they are dry. Wet weather promotes growth and sporulation of the fungus and allows the dark pigment to move from the pod to the seed coat. If harvested seeds are too moist (>18% water content), the moisture in the seeds promotes growth and sporulation of the fungus during handling and storage, particularly under warm and humid conditions. The seeds should be dried as soon as possible if the moisture content is high, and seed lots with different moisture contents should not be mixed. The fungus may grow in pockets of higher moisture content during storage.

Fig. 35. Alternaria pod spot, caused by *Alternaria alternata* (left). Healthy pods are on the right. (Courtesy J. C. Tu)

Some fungicides, such as chlorothalonil, are not only ineffective against *A. alternata* but can significantly increase the incidence of black pod. The fungus is, however, sensitive to iprodione.

Selected References

Abawi, G. S., Crosier, D. C., and Cobb, A. C. 1977. Pod-flecking of snap beans caused by *Alternaria alternata*. Plant Dis. Rep. 61:901-905.

Schwartz, H. F. 1989. Additional fungal pathogens. Pages 231-259 in: Bean Production Problems in the Tropics, 2nd ed. H. F. Schwartz and M. A. Pastor-Corrales, eds. Centro Internacional de Agricultura Tropical (CIAT), Cali, Colombia.

Tu, J. C. 1982. Etiology of seed coat discoloration of white beans. Can. J. Plant Sci. 62:277-284.

Tu, J. C. 1983. Efficacy of iprodione against Alternaria black pod and white mold of white beans. Can. J. Plant Pathol. 5:133-135.

Tu, J. C. 1985. Biology of *Alternaria alternata*, the causal agent of black pod disease of white bean in southwestern Ontario. Can. J. Plant Sci. 65:913-919.

(Prepared by J. C. Tu)

Angular Leaf Spot

Angular leaf spot is a major bean disease in tropical and subtropical areas. The disease affects most aerial parts of the plant, especially the foliage and pods in the field throughout the growing season. It is particularly destructive at or after flowering and in areas where warm, moist conditions are accompanied by abundant inoculum from infested plant residues and contaminated seeds. Angular leaf spot causes premature leaf drop and foliar and stem necrosis that culminate in poorly filled seeds and reduced seed quality. Yield losses can reach 80% when susceptible cultivars are grown and environmental conditions are conducive for disease development.

Symptoms

Symptoms occur on all aerial plant parts, but lesions are most characteristic on leaves 8–12 days after infection. Leaf lesions initially appear as gray or brown irregular spots that may be bordered by a chlorotic halo. By 9 days after infection, the lesions become necrotic and assume the angular shape characteristic of the disease (Fig. 36). More-severe symptoms include coalescence of lesions and chlorosis, necrosis, and premature abscission of affected leaves. On primary leaves, lesions are generally circular rather than angular in shape and may develop concentric rings. Pod lesions are large, oval to circular, reddish brown spots, usually surrounded by a darker-colored border (Fig. 37). Pod lesions can be easily confused with those caused by halo and common bacterial blights, especially as the crop matures. Lesions on stems and petioles are dark brown and elongate. Black synnemata and conidia are produced in lesions on the lower surface of infected leaves, pods, stems, and petioles, and sporulation occurs only after 24–48 h of continuous humid or moist conditions.

Causal Organism

Angular leaf spot is caused by *Phaeoisariopsis griseola* (Sacc.) Ferraris, an imperfect fungus belonging to the order Moniliales. Conidiophores are produced in groups (synnemata) of 8–40 and bear, at the tips, conidia that are pale gray, are cylindrical to spindle shaped, may be slightly curved, measure $3-9 \times 34-83$ μm, and have zero to six septa. Considerable variation in conidial size and septation occurs both between and within isolates. Sporulation occurs at 16–26°C. The fungus infects numerous crops, including common bean, lima bean, scarlet runner bean, tepary bean, black gram or mung bean, pea, cowpea, *Vigna angularis* (Willd.) Ohwi & H. Ohashi, *V. umbellata* (Thunb.) Ohwi & H. Ohashi, soybean, and *Lablab purpureus* (L.) Sweet subsp. *purpureus.*

Disease Cycle and Epidemiology

Sources of primary inoculum include volunteer plants, off-season crops, contaminated seeds, and infested plant residue. Spores produced in seeds or other host tissue are disseminated to leaves by wind, rain splash, or both.

Conidia germinate in the presence of water or high humidity and enter the host through stomata. Growth continues intercellularly in the mesophyll and palisade layers, resulting in tissue disintegration that extends to the upper epidermis. Later, the fungus grows intercellularly in the necrotic tissues, becoming delimited by the vascular bundles in the veins. By 9 days after infection, the pathogen extensively colonizes necrotic lesions. After 9–12 days, stromata develop in substomatal cavities, synnemata form, and sporulation occurs during periods of high humidity. These spores cause secondary spread of the disease. A minimum of 3 h of high humidity (95–100%) is sufficient for infection to take place, although increasing this to 24 h increases the rate of infection. Although infection and

Fig. 37. Pod and branches with symptoms of angular leaf spot, caused by *Phaeoisariopsis griseola.* (Courtesy H. F. Schwartz, from the files of G. Alvarez)

Fig. 36. Angular leaf spot, caused by *Phaeoisariopsis griseola.* (Courtesy H. F. Schwartz)

disease development occur over a range of temperatures (16 to 28°C), the optimum temperature for disease initiation and further development of the disease is 24°C. Infection decreases above and below 24°C, ceasing above 36°C and below 5°C. The incubation period at 24°C is about 5–7 days, and this period extends to 15 days under cooler conditions (16°C).

Once infection occurs, the pathogen can develop successfully and form numerous stromata under relatively dry conditions. However, epidemic development is most rapid under conditions of high relative humidity and moderate temperatures alternating with periods of wind and low humidity. Symptoms in the field are generally observed soon after flowering or as plants approach maturity. Crop losses result primarily from premature defoliation. Seeds may be infected or infested. Infestation in red kidney bean is associated with fungal development in the hilum area only, whereas in other beans, the fungal development can be in the hilum and other areas of the seed coat. Levels of seed infection differ among bean cultivars. The viability of *P. griseola* in seeds decreases over time.

Management

High levels of resistance to angular leaf spot are available in some common bean genotypes but may be affected by the variability of pathogen races in different production regions. One or more independent dominant or independent recessive genes, either alone or in combination, depending on the host–pathogen interaction, confer the resistance of some genotypes. Partial resistance or quantitative traits against angular leaf spot are common in beans, but these are managed by an undetermined number of genes. Advances in the understanding of the evolution of the pathogen have shown that resistance genes from the Mesoamerican gene pool are highly effective against Andean isolates of *P. griseola* and vice versa. High levels of resistance have also been identified in the secondary *Phaseolus* genotypes composed of *Phaseolus coccineus* L. and *Phaseolus polyanthus* Greenman. Many highly resistant cultivars, exhibiting both quantitative and qualitative resistance, are available and should be used as the main component in an integrated management approach.

In tropical and subtropical areas where conditions are conducive for disease development all year, an integrated management approach that combines host resistance with other means of angular leaf spot management should be adopted. Pathogen-free seeds that have been treated with an effective fungicide should be planted. Rotation with a nonhost crop should be practiced where possible, with a break of 2 years between bean crops to permit decomposition of infested residue. Infested crop residue should be removed or buried and destroyed by deep plowing or other means. Planting beans in sites adjacent to fields in which beans have recently been harvested should be avoided. Fungicides should be applied as foliar sprays when disease is first detected and if environmental conditions are favorable for disease spread. Effective fungicides include bitertanol, captafol, chlorothalonil, maneb, mancozeb, metiram, and zineb. The optimum time of initial application and the number and frequency of sprays varies with chemical and prevailing environmental conditions.

Selected References

Allen, D. J., Buruchara, R. A., and Amithson, J. B. 1997. Diseases of common bean. Pages 179-265 in: The Pathology of Food and Pasture Legumes. D. J. Allen and J. M. Lenné, eds. CAB International, Wallingford, U.K.

Correa-Victoria, F. J., Pastor-Corrales, M. A., and Saettler, A. W. 1989. Angular leaf spot. Pages 59-75 in: Bean Production Problems in the Tropics, 2nd ed. H. F. Schwartz and M. A. Pastor-Corrales, eds. Centro Internacional de Agricultura Tropical (CIAT), Cali, Colombia.

Guzmán, P., Gilbertson, R. L., Nodari, R., Johnson, W. C., Temple, S. R., Mandala, D., Mkandawire, A. B. C., and Gepts, P. 1995. Characterization of variability in the fungus *Phaeoisariopsis griseola* suggests coevolution with the common bean (*Phaseolus vulgaris*). Phytopathology 85:600-607.

Liebenberg, M. M., and Pretorius, Z. A. 1997. A review of angular leaf spot of common bean (*Phaseolus vulgaris* L.). Afr. Plant Prot. 3:81-106.

Schwartz, H. F., Pastor-Corrales, M. A., and Singh, S. P. 1982. New sources of resistance to anthracnose and angular leaf spot of beans (*Phaseolus vulgaris* L.). Euphytica 31:741-754.

Sindhan, G. S., and Bose, S. K. 1980. Epidemiology of angular leaf spot of French bean caused by *Phaeoisariopsis griseola*. Indian Phytopathol. 33:64-68.

(Prepared by A. W. Saettler; Revised by G. Mahuku)

Anthracnose

Anthracnose is distributed worldwide but causes greater losses in temperate and subtropical zones than in the tropics. It occurs in North, Central, and South America, as well as in Europe, Africa, Australia, and Asia. Anthracnose is one of the most important diseases of beans throughout the world, and yield losses can reach 100% when contaminated seeds are planted and prolonged conditions favorable to disease development occur during the crop cycle.

Symptoms

The disease can affect all aerial parts of the bean plant. Infected cotyledons exhibit small, dark brown to black lesions. Conidia and hyphae may be transported by rain or dew to the developing hypocotyl, where infected tissue develops minute, rust-colored specks that enlarge longitudinally to form sunken lesions or eyespots. These may cause a rotting of the hypocotyl. On older stems, lesions may reach 5–7 mm in length.

Lesions are more common on leaf petioles and on the lower surface of leaves and leaf veins. They are elongate, angular, and brick red to purple, becoming dark brown to black. Lesions

Fig. 38. Anthracnose, caused by *Glomerella lindemuthiana* (anamorph *Colletotrichum lindemuthianum*), on a leaf. (Courtesy H. F. Schwartz)

may also appear on the upper surface of leaves (Fig. 38). Pod infections appear as tan to rust-colored lesions that develop into sunken cankers (1–10 mm in diameter) delimited by a slightly raised black ring surrounded by a reddish brown border (Fig. 39). Young pods may shrivel and dry if severely infected. The fungus can invade the pod and infect the cotyledons and seed coat of developing seeds. Infected seeds are often discolored and may contain dark brown to black cankers (Fig. 40).

Sporulation can occur in lesions on the petioles, large leaf veins, and pods. The centers of these lesions are light colored and, during periods of low temperatures and high moisture levels, may contain a gelatinous mass of tan conidia that dry to gray-brown or black granules.

Causal Organism

Anthracnose is caused by the fungus *Glomerella lindemuthiana* Shear (teleomorph), more commonly seen on beans in its anamorph, *Colletotrichum lindemuthianum* (Sacc. & Magnus) Lams.-Scrib. The fungus is pathogenic to many legume species, including common bean, scarlet runner bean, black gram, mung bean, cowpea, and fava bean. Hyphae are hyaline to gray

Fig. 39. Anthracnose, caused by *Glomerella lindemuthiana* (anamorph *Colletotrichum lindemuthianum*), on pods and seeds. (Courtesy H. F. Schwartz)

Fig. 40. Anthracnose, caused by *Glomerella lindemuthiana* (anamorph *Colletotrichum lindemuthianum*), on seeds (right). Healthy seeds are on the left. (Courtesy R. Hall)

at first, rapidly becoming dark to nearly black, with compact aerial mycelium upon maturity. Maximum growth in culture occurs at 22–24°C. Conidia are borne on an acervulus. The rounded or elongate acervuli are intraepidermal and subepidermal and may reach 300 μm in diameter. Setae may form at the margin of an acervulus or in culture, are brown and septate, and measure 4–9 × 100 μm. Pale, tan to salmon-colored spore masses form in acervuli. Conidia are unicellular, hyaline, cylindrical (with both ends obtuse or with a narrow and truncate base), and uninucleate; they generally have a clear, vacuolelike body near the center. Conidia measure 2.5–5.5 × 9.5–22 μm.

Disease Cycle and Epidemiology

The fungus survives between crops in crop residue and can be disseminated in seeds, air, and water. Conidia that reach plant surfaces may germinate in 6–9 h under favorable environmental conditions to form one to four germ tubes and appressoria that attach to the host cuticle by a gelatinous layer. The infective hypha penetrates the cuticle and epidermis by mechanical pressure as it grows from the appressorium. Infective hyphae enlarge and grow between the cell wall and the protoplast for 2–4 days without apparent damage to host cells. Several days later, cell walls are degraded enzymatically, leading to the appearance of water-soaked lesions that darken because of a high content of tannins. Mycelium aggregates within the lesion site and forms an acervulus that ruptures the host cuticle. Conidia then form within a water-soluble gelatinous matrix and serve as secondary inoculum.

Conidial production and plant infection are favored by temperatures of 13–26°C, with an optimum of 17°C. Relative humidity greater than 92% (or free moisture) is required during all stages of conidium germination, incubation, and subsequent sporulation. Moderate rainfalls at frequent intervals, particularly if accompanied by wind or splashing rain, are essential for local dissemination of conidia and development of severe epidemics.

Management

Certain cultivars of beans are resistant to infection by *C. lindemuthianum*. However, the pathogen is extremely variable pathogenically, especially when isolates from South America, Central America, Europe, and Africa are compared.

Seed coat infestations are managed with various chemical treatments, including thiophanate methyl. Preventive spraying with protectant or systemic fungicides, such as thiophanate methyl, captafol, chlorothalonil, and carbendazim, has limited effectiveness. If applications are made, they should coincide with flower initiation, late flowering, and pod fill if disease pressure is high and environmental conditions favor infection.

Production of pathogen-free seeds in areas with dry growing seasons has been successfully used to reduce disease losses in the United States and elsewhere. Rotation for 2–3 years with nonhost crops, such as cereals and corn, may reduce disease severity by reducing initial inoculum levels from infested debris.

Selected References

Allen, D. J., and Lenné, J. M. 1998. Anthracnose. Pages 182-192 in: The Pathology of Food and Pasture Legumes. CAB International, Wallingford, U.K.

Balardin, R. S., and Kelly, J. D. 1998. Interaction between races of *Colletotrichum lindemuthianum* and gene pool diversity in *Phaseolus vulgaris* L. J. Am. Soc. Hortic. Sci. 123:1038-1047.

Balardin, R. S., Jarosz, A. M., and Kelly, J. D. 1997. Virulence and molecular diversity in *Colletotrichum lindemuthianum* from South, Central, and North America. Phytopathology 87:1184-1191.

Holliday, P. 1980. Fungus Diseases of Tropical Crops. Cambridge University Press, New York.

Kelly, J. D., and Vallejo, V. A. 2004. A comprehensive review of the major genes conditioning resistance to anthracnose in common bean. HortScience 39:1196-1207.

Mahuku, G. S., Jara, C. E., Cajiao, C., and Beebe, S. 2002. Sources of resistance to *Colletotrichum lindemuthianum* in the secondary gene pool of *Phaseolus vulgaris* and in crosses of primary and secondary gene pools. Plant Dis. 86:1383-1387.

Pastor-Corrales, M. A., and Tu, J. C. 1989. Anthracnose. Pages 77-104 in: Bean Production Problems in the Tropics, 2nd ed. H. F. Schwartz and M. A. Pastor-Corrales, eds. Centro Internacional de Agricultura Tropical (CIAT), Cali, Colombia.

(Prepared by H. F. Schwartz;
Revised by M. A. Pastor-Corrales)

Ascochyta Leaf Spot

Ascochyta (Phoma) leaf spot occurs in Latin America, western Europe, eastern Africa, Australia, and the United States. It is economically important only in cool, humid regions. It can affect all aerial parts of beans, and the causal fungus can be seedborne. Yield losses can exceed 40%.

Symptoms

Infected leaves exhibit brown to black lesions (Fig. 41) that may develop concentric zones 10–30 mm in diameter (Fig. 42) and may also contain small, black pycnidia. Concentric dark gray to black lesions also may appear on branches, stems, nodes, and pods (Fig. 43) and cause stem girdling and plant death. Infected seeds become brown to black (Fig. 44). The fungus may spread systemically throughout the plant, and defoliation and pod drop may also occur.

Causal Organisms

Ascochyta leaf spot and blight (speckle disease) is caused by *Phoma exigua* Desmaz. var. *exigua* Desmaz., formerly called *Ascochyta phaseolorum* Sacc. *P. exigua* Desmaz. var. *diversispora* (Bubak) Boerema (syn. *P. diversispora* Bubak) is also reported as a pathogen causing Ascochyta blight (black node) of beans in western Europe and eastern Africa. A third species, *A. boltshauseri* Sacc. in Boltshauser (syn. *Stagonosporopsis hortensis* (Sacc. & Malbr.) Petr.), causes reddish brown leaf

Fig. 43. Ascochyta leaf spot, caused by *Phoma exigua* var. *exigua*, showing a zonate lesion on a pod. (Courtesy H. F. Schwartz, from the files of M. A. Pastor-Corrales)

Fig. 41. Ascochyta leaf spot, caused by *Phoma exigua* var. *exigua*. (Courtesy H. F. Schwartz)

Fig. 42. Ascochyta leaf spot, caused by *Phoma exigua* var. *exigua*, showing zonate leaf spots. (Courtesy H. F. Schwartz, from the files of M. A. Pastor-Corrales)

Fig. 44. Ascochyta leaf spot, caused by *Phoma exigua* var. *exigua*, showing discoloration of a seed. (Courtesy H. F. Schwartz, from the files of M. A. Pastor-Corrales)

spots (speckle) on beans. *Phoma* spp. produce hyaline, septate, submerged mycelium in culture and conidia that are usually two-celled and measure 5 × 20 μm. Pycnidia are 220 μm in diameter. In contrast, pycnidia of *P. exigua* var. *diversispora* and *P. exigua* var. *exigua* average 150 μm in diameter, and the cylindrical to oval, allantoid, hyaline, pale yellowish brown conidia are usually one-celled and measure 2–3 × 5–10 μm.

Disease Cycle and Epidemiology

The fungus survives in crop debris and seeds. Infection of the plant by conidia is favored by high humidity, frequent rains accompanied by wind, and temperatures below than 28°C. Sporulation and germination are optimal at 21°C, and mycelial growth is optimal at 24°C.

Management

Sprays with fungicides, such as chlorothalonil and carbendazim, may reduce disease pressure and yield losses. Sanitation and rotation with nonhost crops, such as cereals and corn, may reduce disease severity by reducing initial inoculum levels. Treating seeds with fungicides and using seeds free from the pathogen may help. Partial resistance has been identified in common bean in eastern Africa and in *Phaseolus coccineus* L.

Selected References

Alcorn, J. L. 1968. Occurrence and host range of *Ascochyta phaseolorum* in Queensland. Aust. J. Biol. Sci. 21:1143-1151.

Allen, D. J., and Lenné, J. M. 1998. Ascochyta blight. Pages 198-201 in: The Pathology of Food and Pasture Legumes. CAB International, Wallingford, U.K.

Boerema, G. H., Cruger, G., Gerlagh, M., and Nirenberg, H. 1981. *Phoma exigua* var. *diversispora* and related fungi on *Phaseolus* beans. J. Plant Dis. Prot. 88:597-607.

Schwartz, H. F. 1989. Additional fungal pathogens. Pages 231-259 in: Bean Production Problems in the Tropics, 2nd ed. H. F. Schwartz and M. A. Pastor-Corrales, eds. Centro Internacional de Agricultura Tropical (CIAT), Cali, Colombia.

Sutton, B. C., and Waterston, J. M. 1966. *Ascochyta phaseolorum*. Descriptions of Pathogenic Fungi and Bacteria, No. 81. Commonwealth Mycological Institute and Association of Applied Biologists, Kew, Surrey, England.

(Prepared by H. F. Schwartz)

Ashy Stem Blight

Ashy stem blight of beans is an economically important disease in many parts of the world, especially in the warmer areas of bean production. It is often a major problem in the southern United States, the Caribbean, and Central and South America. Although it has been known since 1905, it continues to be troublesome because management measures are not always completely effective.

Symptoms

The initial symptom appears as a small, irregularly shaped, blackish, sunken lesion on the seedling stem at the soil line before or soon after emergence (Fig. 45). From this original canker, the infection spreads, especially upward, and several sunken cankers may enlarge, coalesce, and eventually girdle and kill the plant (Fig. 46). These cankers have a definite margin and commonly contain concentric rings. Wilting, chlorosis, and death of leaves may be more pronounced on one side of the plant. Numerous small, black sclerotial bodies or pycnidia form on the aging, ashen gray cankers (Fig. 47). Ashen gray lesions also form on pods (Fig. 48) and seeds (Fig. 49). Large areas of the crop may be killed, especially during pod fill (Fig. 50).

Causal Organism

Ashy stem blight is caused by the fungus *Macrophomina phaseolina* (Tassi) Goidanich. It is a highly variable organism, and nonpycnidial strains are common. The sclerotia are smooth, black, and spherical (50–150 μm in diameter). Sclerotia and pycnidia form just beneath the surface of the lesion and are commonly intermixed. Pycnidia have no stroma. Conidia are nonseptate, hyaline, and 6–12 × 12–34 μm. They are straight or slightly curved, are rounded at one end and pointed at the other, and are borne on short, nearly straight conidiophores.

Disease Cycle and Epidemiology

M. phaseolina contaminates seeds and infests soil. It can be spread by any means that moves infested soil (e.g., farm implements, irrigation water, animals, and wind). In addition, seeds produced in regions where the disease occurs are often contaminated, but seeds grown in seed-producing areas of the western United States appear to be free of the pathogen. The

Fig. 45. Ashy stem blight, caused by *Macrophomina phaseolina*, on the stem of a seedling. (Courtesy H. F. Schwartz)

Fig. 46. Ashy stem blight, caused by *Macrophomina phaseolina*, on a stem. (Courtesy H. F. Schwartz)

Fig. 47. Ashy stem blight, caused by *Macrophomina phaseolina*, on a stem. (Courtesy H. F. Schwartz)

fungus overwinters as sclerotia or pycnidia in debris in infested soil or with the seeds. The fungus may occur beneath the seed coat. Airborne conidia can initiate leaf spots on older plants. The fungus has a wide host range.

The fungus is favored by temperatures above 27°C. Ordinary soil moisture conditions permit disease initiation and development, but high humidity is conducive to infection by conidia dispersed in the air.

Fig. 48. Ashy stem blight, caused by *Macrophomina phaseolina*, on pods. (Courtesy H. F. Schwartz, from the files of M. A. Pastor-Corrales)

Fig. 49. Seeds infected with *Macrophomina phaseolina*. (Courtesy H. F. Schwartz, from the files of M. A. Pastor-Corrales)

Fig. 50. Ashy stem blight, caused by *Macrophomina phaseolina*, in a bean crop. (Courtesy H. F. Schwartz, from the files of M. A. Pastor-Corrales)

Management

Where available, pathogen-free seeds, resistant bean cultivars, chemical treatment of seeds or soil (e.g., with fumigants), and chemical plant protectants are used to manage the disease. Deep plowing and incorporation of organic soil amendments have also been recommended.

Selected References

Echavez-Badel, R., and Beaver, J. S. 1987. Dry bean genotypes and *Macrophomina phaseolina* (Tassi) Goid in inoculated and non-inoculated field plots. J. Agric. Univ. P. R. 71:385-390.

Echavez-Badel, R., and Beaver, J. S. 1987. Resistance and susceptibility of beans, *Phaseolus vulgaris* L., to ashy stem blight, *Macrophomina phaseolina* (Tassi) Goid. J. Agric. Univ. P. R. 71:403-405.

Schwartz, H. F. 1989. Additional fungal pathogens. Pages 231-259 in: Bean Production Problems in the Tropics, 2nd ed. H. F. Schwartz and M. A. Pastor-Corrales, eds. Centro Internacional de Agricultura Tropical (CIAT), Cali, Colombia.

(Prepared by D. J. Hagedorn; Revised by H. F. Schwartz)

Cercospora Leaf Spot and Blotch

Cercospora leaf spot and blotch occurs in Latin America and in the southern United States. It can affect all aerial parts of beans, including seeds; and defoliation has occurred in Colombia. Serious yield losses to beans have not been reported. Other hosts could include lima bean, hyacinth bean, azuki bean, and cowpea.

Symptoms

Infected leaves, especially those that are more mature, exhibit brown or rust-colored lesions that vary in shape (circular to angular) and diameter (2–10 mm) and may coalesce (Fig. 51). Lesions may have a gray center with a slightly reddish border. Conidia develop in this center on short conidiophores. Severely affected leaves become chlorotic (Fig. 52). Lesions may dry and portions may fall out, giving the leaf a ragged appearance. Some species of the fungus cause numerous lesions on primary leaves but seldom infect the trifoliolate leaves. Lesions and blemishes may occur on branches, stems, and pods.

Causal Organisms

Cercospora leaf spot and blotch is primarily caused by *Cercospora canescens* Ellis & G. Martin and *C. cruenta* Sacc. (syn. *Pseudocercospora cruenta* (Sacc.) Deighton). The latter fungus is the imperfect state of *Mycosphaerella cruenta* D. H. Latham. *C. phaseoli* Dearn. & Barth. and *C. caracallae* (Speg.)

Fig. 51. Cercospora leaf spot and blotch, caused by *Cercospora canescens*. (Courtesy H. F. Schwartz)

Vassiljevsky & Karakulin also cause leaf spots of beans. Stromata form in stomatal openings, mostly on the lower surface. Conidiophores, formed in dense fascicles, are pale olivaceous to brown, straight, and geniculate; have up to two septa; and measure 3–6.5 × 20–175 μm. Conidia are pale brown, obclavate, straight or variously curved, multiseptate, and 2.5–5 × 30–300 μm.

Disease Cycle and Epidemiology

Little is known regarding the disease cycle of these pathogens. The fungus contaminates seeds and apparently survives in crop debris. Primary leaves may become infected by *C. cruenta*, a species that seldom affects trifoliolate leaves. However, lesions on primary leaves may serve as a source of infection for trifoliolate leaves after they have matured. Vigorously growing leaves rarely become infected. Conidia form most abundantly at 28°C, and light favors sporulation on an organic culture medium, such as carrot.

Management

Certain cultivars of beans are resistant to infection by *C. canescens*. Chemical sprays are seldom warranted, but copper fungicides are effective if applied early in the disease cycle. Use of clean seeds, sanitation, and rotation with nonhost crops, such as cereals and corn, may reduce initial inoculum levels.

Selected References

Holliday, P. 1980. Fungus Diseases of Tropical Crops. Cambridge University Press, New York.
Mulder, J. L., and Holliday, P. 1975. *Cercospora canescens*. Descriptions of Pathogenic Fungi and Bacteria, No. 462. Commonwealth Mycological Institute and Association of Applied Biologists, Kew, Surrey, England.
Mulder, J. L., and Holliday, P. 1975. *Cercospora cruenta*. Descriptions of Pathogenic Fungi and Bacteria, No. 463. Commonwealth Mycological Institute and Association of Applied Biologists, Kew, Surrey, England.
Schwartz, H. F. 1989. Additional fungal pathogens. Pages 231-259 in: Bean Production Problems in the Tropics, 2nd ed. H. F. Schwartz and M. A. Pastor-Corrales, eds. Centro Internacional de Agricultura Tropical (CIAT), Cali, Colombia.
Sivanesan, A. 1990. *Mycosphaerella cruenta*. Descriptions of Pathogenic Fungi and Bacteria, No. 985. Commonwealth Mycological Institute and Association of Applied Biologists, Kew, Surrey, England.

(Prepared by H. F. Schwartz)

Chaetoseptoria Leaf Spot

Chaetoseptoria leaf spot primarily affects foliage of beans. It occurs in Mexico, Panama, Central America, Venezuela, and the West Indies. The causal fungus has a wide host range within the family Leguminosae and may cause complete defoliation and up to 50% reduction in yield.

Symptoms

Infected leaves exhibit medium to large (10 mm), irregular to circular, ash gray, faintly zonate lesions with a reddish border and gray to black pycnidia in the center (Fig. 53). The disease typically occurs on leaves. No pod or stem infections have been reported. However, seed transmission is suspected to occur in Mexico.

Causal Organism

Chaetoseptoria leaf spot is caused by *Chaetoseptoria wellmanii* J. A. Stevenson. Pycnidia are amphigenous, dark gray, subpyriform, and 120–350 μm in diameter. The ostiole of the pycnidium is well defined, circular, and 15–50 μm in diameter. Setae are straight, have three to nine septa, and measure 3–6 × 60–225 μm. The hyaline conidia are 2.5–4.0 × 75–160 μm, acicular, and straight to curved to flexuous, with six to eight septa.

Disease Cycle and Epidemiology

Primary leaves become infected by conidia from infested debris or soil. Developed, sporulating lesions may serve as a source of conidia for infection of trifoliolate leaves. Infection is favored by cool, moist conditions.

Management

Rotation with nonhost crops, such as cereals and corn, for prolonged periods (5 years in Mexico) may reduce disease severity by reducing initial inoculum levels. Chemical sprays are seldom warranted, but copper fungicides may provide management.

Selected References

Crispin, A., Sifuentes, J. A., and Campos, J. 1976. Enfermedades y plagas del frijol en Mexico. Foll. Divulg. 39. Inst. Nacl. Invest. Agrar., Secretaria de Agricola y Ganadero (SAG).
Punithalingam, E. 1985. *Chaetoseptoria wellmanii*. Descriptions of Pathogenic Fungi and Bacteria, No. 822. Commonwealth Mycologi-

Fig. 52. Cercospora leaf spot and blotch, caused by *Cercospora canescens*. (Courtesy H. F. Schwartz)

Fig. 53. Chaetoseptoria leaf spot, caused by *Chaetoseptoria wellmanii*. (Courtesy H. F. Schwartz, from the files of G. E. Galvez)

cal Institute and Association of Applied Biologists, Kew, Surrey, England.

Schwartz, H. F. 1989. Additional fungal pathogens. Pages 231-259 in: Bean Production Problems in the Tropics, 2nd ed. H. F. Schwartz and M. A. Pastor-Corrales, eds. Centro Internacional de Agricultura Tropical (CIAT), Cali, Colombia.

Yerkes, W. D., Jr. 1956. *Chaetoseptoria wellmanii* in Mexico. Mycologia 48:738-740.

(Prepared by H. F. Schwartz)

Diaporthe Pod Blight

Diaporthe pod blight in beans has been reported from Honduras, but yield loss estimates are not available. Other hosts include cowpea and soybean.

Symptoms

Symptoms appear first on leaves as irregularly shaped, brown lesions surrounded by a distinct border. Black pycnidia and, occasionally, perithecia form in a zone or are scattered throughout the lesions. Pod infections may then occur and pods become discolored with pycnidia present in the lesions. The fungus can be seedborne in soybean and beans.

Causal Organism

Diaporthe pod blight is caused by *Diaporthe phaseolorum* (Cooke & Ellis) Sacc. It produces hyaline, oblong ascospores measuring 2–4 × 10–12 μm and having one septum. The ascospores are produced inside black perithecia, which measure 300 μm in diameter. Pycnidiospores are produced in the black pycnidia, and the oval spores measure 2–5 × 6–9 μm.

Disease Cycle and Epidemiology

Nothing is reported on conditions that favor the disease.

Management

Strategies to manage the disease include crop rotation, planting clean seeds, and using foliar fungicides. Bean germ plasm should be screened to identify sources of resistance, which does occur in soybean.

Selected References

Allen, D. J., and Lenné, J. M. 1998. The Pathology of Food and Pasture Legumes. CAB International, Wallingford, U.K.

Schwartz, H. F. 1989. Additional fungal pathogens. Pages 231-259 in: Bean Production Problems in the Tropics, 2nd ed. H. F. Schwartz and M. A. Pastor-Corrales, eds. Centro Internacional de Agricultura Tropical (CIAT), Cali, Colombia.

Weber, G. F. 1973. Bacterial and Fungal Diseases of Plants in the Tropics. University of Florida Press, Gainesville.

(Prepared by H. F. Schwartz)

Downy Mildew

Downy mildew can affect all aerial parts of beans. It occurs in Central and South America and the West Indies. A related fungus that infects lima bean occurs in the United States, Italy, Russia, Sri Lanka, the former Zaire, and the Philippines. Losses are sporadic and seldom severe.

Symptoms

The pathogen may cause seedling damping-off. On leaves and petioles, lesions appear as white spots that enlarge and eventually cause leaves to wilt and die. Blossoms, buds, and other plant parts may also be killed. Major damage occurs as pods become infected and covered by white, cottony patches of mycelium (Fig. 54). A reddish brown border may develop around this infected portion. Pods are killed, become black and dry, and often remain attached. Mycelium may penetrate from pod to seed tissues.

Causal Organism

Downy mildew is caused by *Phytophthora nicotianae* Breda de Haan var. *parasitica* (Dastur) G. M. Waterhouse and *P. phaseoli* Thaxt. The latter pathogen is primarily associated with downy mildew of lima bean. In both fungi, the mycelium is hyaline, branched, and nonseptate and produces haustoria.

Sporangia of *P. nicotianae* var. *parasitica* are broadly ovoid or obpyriform to spherical and 30–50 μm in diameter. Yellowish brown chlamydospores up to 60 μm in diameter may form within 1–2 weeks in culture. Oogonia average 24–31 μm in diameter and become yellow-brown with age. Antheridia are spherical or oval and measure 10 × 12 μm. Oospores are aplerotic and 18–20 μm in diameter. The fungus grows most rapidly in culture at 30–32°C.

P. phaseoli grows slowly in culture with an optimum temperature of 15–20°C. Sporangia are 22 × 50 μm, oogonia may reach 38 μm in diameter, and oospores and antheridia are 10 × 17 μm. Oospores occur in the host and are abundant in infected pods.

Disease Cycle and Epidemiology

Infection is most serious in pods lying close to the soil and is favored by high moisture levels and low temperatures (especially for *P. phaseoli*). The optimum temperature for zoospore germination and infection is 20°C.

Management

Several lima bean cultivars are resistant to *P. phaseoli*. No dry bean or snap bean cultivar has been reported to have resistance to *P. nicotianae* var. *parasitica*. A fungicide spray applied during flowering and pod formation can reduce pod infection severity. Rotation with nonhost crops, such as cereals and corn, may reduce disease by reducing initial inoculum levels.

Fig. 54. Downy mildew, caused by *Phytophthora nicotianae* var. *parasitica*, on pods. (Courtesy H. F. Schwartz)

Selected References

Evans, T. A., Davidson, C. R., Dominiak, J. D., Mulrooney, R. P., Carrol, R. B., and Antonius, S. H. 2002. Two new races of *Phytophthora phaseoli* from lima bean in Delaware. Plant Dis. 86:813.

Schwartz, H. F. 1989. Additional fungal pathogens. Pages 231-259 in: Bean Production Problems in the Tropics, 2nd ed. H. F. Schwartz and M. A. Pastor-Corrales, eds. Centro Internacional de Agricultura Tropical (CIAT), Cali, Colombia.

Waterhouse, G. M., and Waterston, J. M. 1964. *Phytophthora nicotianae* var. *parasitica*. Descriptions of Pathogenic Fungi and Bacteria, No. 35. Commonwealth Mycological Institute and Association of Applied Biologists, Kew, Surrey, England.

(Prepared by H. F. Schwartz)

Entyloma Leaf Smut

Entyloma leaf smut primarily affects leaves. It occurs in various countries of Central America. Yield loss estimates are not available.

Symptoms

Infected leaves exhibit a blister smut that is evident as dark swellings on the upper leaf surface (Fig. 55). Round or oval lesions (sori) first appear water-soaked and then become gray-brown on the upper leaf surface and gray-blue on the lower leaf surface. Lesions may coalesce and be delimited by leaf veinlets.

Causal Organism

Entyloma leaf spot of common bean is caused by an unidentified species of the genus *Entyloma* and could be confused with infection by *Protomycopsis* spp. *E. petuniae* Speg. has been reported in beans in Argentina, and *E. vignae* Batista et al. has been reported in cowpea in Brazil. The swellings or sori are filled with mycelium and teliospores of the fungus. Teliospores are single, free, and terminal or intercalary on hyphal branches. They are thick walled, irregular in shape, hyaline to yellowish or reddish brown, and about 12–18 μm in diameter.

Disease Cycle and Epidemiology

Infection usually occurs only on primary leaves or first and second sets of trifoliolate leaves, but severe infection of 40–60% of the foliage may occur. The fungus apparently survives in previously infected crop debris.

Management

Rotation with nonhost crops, such as cereals and corn, may reduce disease severity by reducing initial inoculum levels. Chemical sprays are seldom warranted, but benzimidazole fungicides are effective.

Selected References

Allen, D. J., and Lenné, J. M. 1998. The Pathology of Food and Pasture Legumes. CAB International, Wallingford, U.K.

Holliday, P. 1980. Fungus Diseases of Tropical Crops. Cambridge University Press, New York.

Schieber, E., and Zentmyer, G. A. 1971. A new bean disease in the Caribbean area. Plant Dis. Rep. 55:207-208.

Schwartz, H. F. 1989. Additional fungal pathogens. Pages 231-259 in: Bean Production Problems in the Tropics, 2nd ed. H. F. Schwartz and M. A. Pastor-Corrales, eds. Centro Internacional de Agricultura Tropical (CIAT), Cali, Colombia.

Vakili, N. G. 1978. Distribution of smut of beans and cowpeas in tropical America and its possible centre of origin. FAO Plant Prot. Bull. 26:19-24.

(Prepared by H. F. Schwartz)

Floury Leaf Spot

Floury leaf spot affects leaves of dry beans and occurs in eastern and central Africa, Europe, Malaysia, Papua New Guinea, South America, and Central America. Floury leaf spot is among the more serious diseases of common bean at relatively high altitudes in the tropics. No estimates of yield losses caused by this disease are available.

Symptoms

Infected leaves exhibit light green to slightly chlorotic, angular to circular lesions (10–15 mm in diameter) on the upper leaf surface. White, floury mats (aggregates) of conidiophores and conidia form on the lower surface of leaves, rarely on the upper surface (Fig. 56). Serious infection may cause defoliation (Fig. 57).

Causal Organism

Floury leaf spot is caused by *Mycovellosiella phaseoli* (Drummond) Deighton (syn. *Ramularia phaseoli* (Drummond) Deighton). Conidiophores, developing from substomatal cavities, are hyaline, smooth septate, branched, 3–8 × 85 μm, flexuous, and slightly to strongly geniculate. They bear conidial scars. Conidia are hyaline, catenulate in branched chains, smooth, and

Fig. 55. Entyloma leaf smut, caused by an *Entyloma* sp. (Courtesy H. F. Schwartz, from the files of G. E. Galvez)

Fig. 56. Floury leaf spot, caused by *Mycovellosiella phaseoli*. (Courtesy H. F. Schwartz)

ellipsoid, with one or no septa, and they measure 4–6 × 7–18 µm. The fungus exhibits slow, cushionlike growth on potato dextrose agar.

Disease Cycle and Epidemiology

Infection generally appears first on older leaves and then progresses to new foliage, especially during periods of high moisture levels and low temperatures. In Colombia, it is more apparent at elevations between 1,500 and 2,000 m above sea level (masl) than at lower elevations. The fungus apparently survives in previously infected crop debris.

Management

Rotation with nonhost crops, such as cereals and corn, may reduce disease severity by reducing initial inoculum levels. Chemical sprays with thiophanate methyl are effective.

Selected References

Cardona-Alvarez, C., and Skiles, R. L. 1958. Floury leaf spot (mancha harinosa) of bean in Colombia. Plant Dis. Rep. 42:778-780.
Deighton, F. C. 1967. Floury leaf spot of French bean caused by *Ramularia phaseoli* (Drummond) comb. nov. Trans. Br. Mycol. Soc. 50:123-127.
Holliday, P. 1980. Fungus Diseases of Tropical Crops. Cambridge University Press, New York.
Ingham, J. 1986. *Mycovellosiella phaseoli*. Descriptions of Pathogenic Fungi and Bacteria, No. 870. Commonwealth Mycological Institute and Association of Applied Biologists, Kew, Surrey, England.

(Prepared by H. F. Schwartz)

Gray Leaf Spot

Gray leaf spot preferentially affects the lower leaves of beans. It has been reported from Central and South America. No estimates of yield losses are available.

Symptoms

Infected leaves exhibit light green to slightly chlorotic, angular lesions (2–5 mm in diameter) on the upper leaf surface, usually delimited by veins and veinlets. Lesions may coalesce and later become covered by the fine, powdery, grayish white growth of the fungus (Fig. 58). A dense gray mat of conidiophores and conidia form on the lower leaf surface. Severe infection may cause defoliation (Fig. 59).

Causal Organisms

Gray leaf spot is caused by *Cercospora vanderysti* P. Henn. and *C. castellanii* Matta & Belliard. The fungus sporulates in effuse, appressed, tomentose, gray layers that contain long, dense, branched, sinuous or flexuous, widely divergent fascicles (4.0–5.5 × 50–300 µm) on the lower leaf surface. On the upper leaf surface, conidiophores form in short, compact, erect, sinuous, slightly branched or unbranched fascicles that are 4.0–5.5 × 30–115 µm. All conidiophores are multiseptate, irregular in width, and not constricted at the septa. Conidia produced on the upper leaf surface are hyaline, cylindrical to obclavate, and straight to distinctly curved or occasionally bent; have zero to three septa (usually one septum); and measure 4–5 × 16–56 µm. Conidia produced on the lower leaf surface are hyaline, cylindrical to clavate, and straight to curved; have up to three septa (usually two septa), and measure 4–5 × 16–38 µm.

Disease Cycle and Epidemiology

Leaves are infected during periods of high moisture levels and low temperatures, especially at altitudes between 1,500 and 2,200 m above sea level (masl) in various Latin American countries. The fungus apparently survives in previously infected crop debris.

Fig. 58. Gray leaf spot, caused by *Cercospora* spp. (Courtesy H. F. Schwartz)

Fig. 57. Floury leaf spot, caused by *Mycovellosiella phaseoli*, and gray leaf spot, caused by *Cercospora* spp. (Courtesy H. F. Schwartz)

Fig. 59. Gray leaf spot, caused by *Cercospora* spp. (Courtesy H. F. Schwartz)

Management

Certain cultivars and breeding lines are resistant to infection. Rotation with nonhost crops, such as cereals and corn, may reduce disease severity by reducing initial inoculum levels. Sprays with cupric hydroxide are effective.

Selected References

Schwartz, H. F. 1989. Additional fungal pathogens. Pages 231-259 in: Bean Production Problems in the Tropics, 2nd ed. H. F. Schwartz and M. A. Pastor-Corrales, eds. Centro Internacional de Agricultura Tropical (CIAT), Cali, Colombia.

Skiles, R. L., and Cardona-Alvarez, C. 1959. Mancha gris, a new leaf disease of bean in Colombia. Phytopathology 49:133-135.

(Prepared by H. F. Schwartz)

Gray Mold

Gray mold is a common disease of green bean in cool and warm climate zones. It affects all parts of the plant, usually starting on senescent organs, such as cotyledons and flowers. Once established, it spreads readily to other tissues in contact with diseased tissues. Colonized senescent flowers remaining on the pod or adhering to it are the major source of inoculum for infection of pods. Gray mold is sometimes severe in marketed green bean if incipient pod lesions escape notice at harvest. In the field, the disease is most prevalent during cool, humid seasons; in crops with dense stands; and in cultivars with a dense bush habit or with flowers that persist on the pod after senescence.

Symptoms

Symptoms are usually first seen on senescent cotyledons or on tissues damaged by frost, hail, wind, sandblast, or machinery. Infection occurs from conidia dispersed by wind or rain, and invasion of senescent tissue is rapid and necrogenous. The lesion is initially dark and water-soaked or translucent, often with characteristic concentric zonation. The pectin in the host is degraded, producing lesions that are flaccid and slimy. If the stem or petiole is girdled, the distal portions of the plant collapse and are rapidly invaded.

The fungus commonly colonizes senescent blossoms and then quickly invades the pod (if the flower remains attached) or any plant part onto which the flower falls. Infection is favored by humid conditions, when a water film persists between the flower and the underlying tissue. On leaves, conspicuous gray to fawn-colored lesions are formed (Fig. 60), often with concentric zonation and a yellowish margin; on stems and petioles, longitudinal brown streaks are formed; and on pods, the lesions are at first water-soaked and then grayish brown with zonation and sunken (Fig. 61). As diseased tissues dry out, characteristic conidiophores and conidia are formed in a gray-brown, powdery mass, and sclerotia may be formed on bulky tissues, such as stems and pods. Under very humid conditions in transit and market packs, conidia are few, but mats of profuse, dirty white mycelium occur, a condition known as nesting.

Causal Organism

Gray mold is caused by the ascomycete fungus *Botryotinia fuckeliana* (de Bary) Whetzel, more familiar as the conidial anamorph *Botrytis cinerea* Pers.:Fr. The mycelium is relatively thick, hyaline, septate, and branched and initially follows the middle lamellae in the bean plant. Microconidia develop as short chains of blastic phialoconidia, 2–3 μm in diameter, embedded in mucilage. They are produced from a compact penicillate cluster of phialides or sometimes from a single phialide on a germ tube. The microconidia function as spermatia. The conidiophore of the *B. cinerea* anamorph is tall, stout, dark (but paler near the apex), and branched once or twice irregularly. On the host, the mature conidiophore is often flattened and twisted at each septum. The conidiogenous branches at the apex of the conidiophore and its branches are inflated ampullae on which the blastic conidia develop in botryose clusters synchronously on fine denticles (Fig. 62). Conidia are smooth, hyaline, aseptate, ovate to elliptical with a slightly protuberant hilum, and 4–11 × 6–18 μm. Sclerotia are formed on fleshy host tissues and are 2–4 mm in diameter. They are plano-convexoid, with

Fig. 61. Gray mold, caused by *Botryotinia fuckeliana* (anamorph *Botrytis cinerea*), on a senescent blossom and a pod. (Courtesy H. F. Schwartz)

Fig. 60. Gray mold, caused by *Botryotinia fuckeliana* (anamorph *Botrytis cinerea*), on a leaf. (Courtesy W. R. Jarvis)

Fig. 62. Conidia and hypha of *Botrytis cinerea*. (Courtesy W. R. Jarvis)

a dark, differentiated rind that is poorly developed at the point of attachment.

Apothecia are rarely reported from the field but may be more common than are generally supposed. They arise from sclerotia, occasionally together with conidiophores (Fig. 63). Apothecia are brown and cupulate, with a slender stipe 2–20 mm long. The disk is infundibuliform to discoid and 1–5 mm in diameter. Asci are unitunicate, inoperculate, long clavate, 10 × 120–150 μm, and eight-spored. Ascospores are uniseriate, hyaline, unicellular, ellipsoidal, and 3.5–4.0 × 8.5–10.0 μm. Asci are interspersed with paraphyses in the hymenium.

Disease Cycle and Epidemiology

Infection occurs within a water film from either ascospores or conidia. Senescent and damaged cotyledons are usually the first organs affected, followed shortly by young stems and leaves. Infected stems are the most durable source of secondary inoculum and produce conidia continuously into the bloom period. Flowers are susceptible, particularly when senescent, and infected flowers constitute the main source of inoculum that infects leaves and pods, either when still attached to the pod or when detached and adhering with a water film. Infection then occurs by direct hyphal growth. Sclerotia, produced on stem and pod tissues, remain in the field with debris or are dispersed in uncleaned seed lots.

B. cinerea is a plurivorous fungus with several hundred hosts among crop plants and weeds. Conidia can be produced on sclerotia and stromata in infected tissues throughout the growing season. It may be assumed that conidia are almost always present in the air. The presence of ascospore inoculum is less certain but has been reported in New York.

Conidia are released by a hygroscopic mechanism that depends on rapidly changing relative humidity. They are disseminated in the air, and most infections occur close to the source of inoculum. Conidia are also dispersed by splashing water. Ascospores are released from apothecia by a puffing mechanism and dispersed in the air.

Both conidia and ascospores require a film of water in which to infect the host, and infection occurs within 20 h at temperatures of 10–25°C, with the optimum near 20°C. Infection is faster in the presence of aphid honeydew, pollen, and other exogenous nutrients. Infection from necrotic, colonized tissue is less dependent on a water film and is very rapid, overcoming resistance in cultivars normally resistant to infection from conidia. Sporulation on diseased tissues occurs 2–3 days after infection and over wide ranges of temperature and humidity. Conidia are produced over a long period from sclerotia lying on the soil or on plant debris. Flowers fallen to the soil are also sources of abundant conidial inoculum after colonization.

Fig. 63. Sclerotium, apothecium, and conidia of *Botryotinia fuckeliana* (anamorph *Botrytis cinerea*). (Courtesy H. F. Schwartz, from the files of G. S. Abawi)

The number of prebloom infections, which largely determines the incidence of pod rot, depends on the number of sporulating sites in the crop, the interval between irrigations, the cumulative duration of leaf wetness, and the canopy size.

Management

Since microclimate (and particularly plant surface wetness) is an important element in the infection process, cultural practices can be manipulated to avoid the disease or reduce its severity. Dense, unthinned rows that are perpendicular to the prevailing wind, overfertilized with nitrogen, overirrigated, or irrigated overhead should be avoided. Weeds should be managed, since they provide alternative sites for sporulation and a favorable microclimate for infection. Transit pod rots and nesting are managed by holding pods at 7–10°C with adequate ventilation.

Resistant cultivars tend to be those that avoid disease. They resist lodging; have an upright, open habit; have small, nonpersistent flowers; and produce pods that do not touch the soil.

Overwintering sclerotia, the main source of initial inoculum, should be plowed under with crop and weed debris. Cereals and corn are rarely hosts for *B. cinerea* and should precede beans in disease-prone areas.

Chemical management is available but rarely considered to be economical for bean crops. If used, fungicides should be applied as soon as flowering begins and applications continued through flowering. Pathogen resistance to certain fungicides is possible and a fungicide resistance management program utilizing fungicide mixtures or alternating between classes of fungicides may need to be implemented.

Selected References

Coley-Smith, J. R., Verhoeff, K., and Jarvis, W. R., eds. 1980. The Biology of *Botrytis*. Academic Press, London.

Jarvis, W. R. 1977. *Botryotinia* and *Botrytis* species: Taxonomy, physiology, and pathogenicity. Monogr. 15. Canadian Department of Agriculture, Ottawa, Ontario, Canada.

Johnson, K. B., and Powelson, M. L. 1983. Influence of prebloom disease establishment by *Botrytis cinerea* and environmental and host factors on gray mold pod rot of snap bean. Plant Dis. 67:1198-1202.

Polach, F. J., and Abawi, G. S. 1975. The occurrence and biology of *Botryotinia fuckeliana* on beans in New York. Phytopathology 65:657-660.

(Prepared by W. R. Jarvis; Revised by R. Forster)

Phyllosticta Leaf Spot

Phyllosticta leaf spot can affect all aerial parts of beans. It occurs throughout Latin America, Europe, Japan, South Africa, and the United States. Yield loss estimates are not available.

Symptoms

Infected leaves, especially those that are mature, exhibit early symptoms of small, angular, water-soaked spots that may coalesce and enlarge to 7–10 mm in diameter. Lesions have a light-colored, necrotic center and are surrounded by a reddish brown margin. The centers of old lesions may fall out, and leaves appear tattered. Small, black, lenticular pycnidia can develop throughout the lesion and along the margin. Lesions may also appear on petioles and stems, and flower buds may become necrotic. Small lesions (1 mm in diameter) with dark centers and reddish margins may develop on pods.

Causal Organism

Phyllosticta leaf spot is caused by *Phyllosticta phaseolina* Sacc. The mycelium is hyaline, branched, septate, and mostly submerged in host tissue. Pycnidia are 70–90 μm in diameter and mostly obovate. Pycnidiospores are hyaline, oval to oblong, one-celled, and $2–3 \times 4–6$ µm.

Disease Cycle and Epidemiology

Leaves become infected during periods of high humidity and rainfall with moderate temperatures. The fungus apparently survives in previously infected crop debris.

Management

Rotation with nonhost crops, such as cereals and corn, may reduce disease severity by reducing initial inoculum levels. Germ plasm should be screened to identify sources of resistance if this pathogen consistently threatens beans. Chemical sprays are seldom warranted, but copper or carbamate fungicides are effective.

Selected References

Goth, R. W., and Zaumeyer, W. J. 1963. Occurrence of Phyllosticta leaf spot in beans in 1963. Plant Dis. Rep. 47:1079.
Schwartz, H. F. 1989. Phyllosticta leaf spot. Page 244 in: Bean Production Problems in the Tropics, 2nd ed. H. F. Schwartz and M. A. Pastor-Corrales, eds. Centro Internacional de Agricultura Tropical (CIAT), Cali, Colombia.

(Prepared by H. F. Schwartz)

Pink Pod Rot

Pink pod rot usually appears on senescent or mature bean plants under prolonged damp conditions. A severe infection was found for the first time in a commercial field planted with cv. Harofleet near Fargo, North Dakota, in late September of 1983.

Symptoms

The disease is characterized by the appearance of a powdery mold that is initially white and eventually pink. On green pods, the infected areas are water-soaked initially, but as the disease progresses, lesions enlarge and the centers become chocolate brown. White, powdery mycelium then develops in the center of the lesions, spreads outward (Fig. 64), and finally turns pink. In maturing pods, the diseased seeds become yellowish brown to chocolate brown and may be shriveled and covered with powdery pink mold (Fig. 65). This development is faster in mature pods than in green pods. Pink mold is also apparent on the dead leaves, petioles, and stems of mature plants.

Causal Organism

Pink pod rot is caused by *Trichothecium roseum* (Pers.:Fr.) Link. The fungus has long, slender, simple, septate conidiophores. Conidia ($8–10 \times 12–18$ µm) are usually borne singly, but they may be held in groups at the apex of conidiophores. Conidia are hyaline, two-celled, and ovoid to ellipsoid. On potato dextrose agar, the colony is initially white, with scant aerial mycelia. Later, the colony surface becomes powdery and yellowish pink.

Disease Cycle and Epidemiology

T. roseum is a common soil fungus that is saprophytic or weakly parasitic and is frequently isolated from seeds of soybean, pea, fava bean, and beans. The fungus apparently survives on plant debris during the winter. Although the biology and ecology of this fungus are not well-known, the epidemiology of pink pod rot is thought to be similar to that of gray mold, except that pink pod rot is usually found on senescent or mature bean plants.

Management

Little information is available on management of pink pod rot. Pathogen-free seeds should be obtained from areas with dry growing seasons. The crop should be harvested as soon as possible, and crop residues should be plowed down.

Selected Reference

Tu, J. C. 1985. Pink pod rot of bean caused by *Trichothecium roseum*. Can. J. Plant Pathol. 7:55-57.

(Prepared by J. C. Tu)

Powdery Mildew

Powdery mildew can affect all aerial parts of beans and is distributed worldwide. The pathogen seldom causes extensive damage, but yield losses up to 69% were reported from Colombia when dry bean infection occurred before flowering.

Fig. 65. Pink pod rot (left), caused by *Trichothecium roseum*. Healthy pods are on the right. (Courtesy J. C. Tu)

Fig. 64. Pink pod rot, caused by *Trichothecium roseum*. (Courtesy J. C. Tu)

Symptoms

Slightly darkened, mottled spots (about 10 mm in diameter) develop on the upper surface of leaves and subsequently become covered by a circular growth of white, superficial, powdery mycelium (Fig. 66). The entire leaf and plant may be covered by cottony fungal growth and may become distorted and yellow (Fig. 67). Premature senescence may occur. Stems and pods can be infected (Fig. 68), leading to additional yield loss. Pods may be stunted, malformed, or killed.

Causal Organism

Powdery mildew is caused by *Erysiphe polygoni* DC.; its conidial anamorph is *Oidium balsamii* Mont. The mycelium is septate, hyaline, white in mass, branched, and supported by numerous hyphae that penetrate the epidermis. The mycelium produces conidiophores that develop hyaline, terminal, one-celled, ellipsoidal to slightly elongate conidia that are often in chains. Conidia are 15–23 × 26–52 μm. Perithecia rarely form on infected beans; they are black, spherical, appendaged, and about 120 μm in diameter. They contain several asci, each with two to four ascospores that are 11–13 × 24–28 μm. The append-ages on the perithecia are numerous, narrow, contorted, simple, hyphalike, and interwoven with mycelium but distinguishable by the light brown coloring at the base, near the attachment to the perithecium.

Disease Cycle and Epidemiology

Conidia dispersed by wind infect plant parts at any stage of plant development. Infection is favored by periods of low humidity and low temperatures. The disease seldom appears until late in the season, when there is little threat to crop productivity. However, infections during flowering and early pod formation may warrant management measures. The fungus can survive on contaminated seeds and debris, especially if perithecia form.

Management

Certain cultivars are resistant to infection by *E. polygoni*. However, since the fungus is pathogenically variable, the resistance may not be stable over time. Rotation with nonhost crops, such as cereals and corn, may reduce disease severity by reducing initial inoculum levels. Chemical sprays are seldom warranted.

Selected References

Schwartz, H. F. 1989. Additional fungal pathogens. Pages 231-259 in: Bean Production Problems in the Tropics, 2nd ed. H. F. Schwartz and M. A. Pastor-Corrales, eds. Centro Internacional de Agricultura Tropical (CIAT), Cali, Colombia.

Schwartz, H. F., Katherman, M. J., and Thung, M. D. T. 1981. Yield response and resistance of dry beans to powdery mildew in Colombia. Plant Dis. 65:737-738.

Zaumeyer, W. J., and Thomas, H. R. 1957. A monographic study of bean diseases and methods for their control. U.S. Dep. Agric. Tech. Bull. 868.

(Prepared by H. F. Schwartz)

Fig. 66. Leaf affected by powdery mildew, caused by *Erysiphe polygoni*. (Courtesy H. F. Schwartz, from the files of S. K. Mohan)

Fig. 67. Plant affected by powdery mildew, caused by *Erysiphe polygoni*. (Courtesy H. F. Schwartz)

Fig. 68. Pods affected by powdery mildew, caused by *Erysiphe polygoni*. (Courtesy H. F. Schwartz)

Rust

Bean rust occurs worldwide but is most common in humid tropical and subtropical areas. In Latin America, it is most serious in Brazil, the Caribbean, Central America, and Mexico. It is also serious in eastern and southern African countries. Severe epidemics occur sporadically in temperate climates, but individual fields may suffer losses more frequently. Bean rust is rare in arid climates, except under irrigation. Yield losses can approach 100% and are directly related to earliness and severity of infection.

Symptoms

Bean rust exhibits typical reddish brown, circular, erumpent uredinial pustules on leaves (Fig. 69), pods (Fig. 70), or stems. The uredinium (pustule) may vary in size from that of a pinpoint to several millimeters in diameter. Initial symptoms appear 5–6 days after infection as minute, whitish, slightly raised spots that develop into pustules several days later. Larger uredinia are often surrounded by a halo of yellow host tissue and occasionally by a ring of smaller secondary uredinia.

Causal Organism

Bean rust is caused by *Uromyces appendiculatus* (Pers.: Pers.) Unger. The prevalent host is common bean, but it has been reported to infect many other *Phaseolus* spp. and a few species of the genus *Vigna*. The primary rust pathogen of *Vigna* spp. is *U. vignae* Barclay. Natural occurrence in lima bean is rare. *U. appendiculatus* has been found in Andean and Middle American wild and weedy *Phaseolus* spp.

Disease Cycle and Epidemiology

U. appendiculatus is an obligate parasite with an autoecious, macrocyclic life cycle. The urediniospores are 18–29 × 20–33 μm, obovoid or broadly ellipsoid, cinnamon or golden brown, and echinulate (Fig. 71). Repeated generations of urediniospore infections occur over most of the growing season. Under appropriate conditions, telia develop within aged uredinia and produce chestnut brown (nearly black), ovoid or ellipsoid to globoid, thick-walled (with walls 2–4 μm) teliospores that are 20–29 × 24–35 μm. Immediately after the teliospores are formed, the two nuclei within the dikaryotic cells fuse to produce a large diploid nucleus. Following a period of dormancy, the teliospores germinate to produce a metabasidium, in which meiosis occurs and on which are produced four binucleate or uninucleate basidiospores. The basidiospores are reniform to ovate-elliptical, smooth, hyaline, and 5.8–11.4 × 10.7–20.0 μm. Basidiospore infection produces spermagonia (pycnia) on the upper (adaxial) leaf surface (Fig. 72). Pycnia appear as chlorotic flecks that enlarge to about 2 mm in diameter and produce white nectar containing the pycniospores. After transfer of pycniospores to a pycnium of the opposite mating type and cross-fertilization, circular clusters (1–2 mm in diameter) of white aecia form on the lower (abaxial) surface of the leaf (Fig. 73). The aecia produce colorless, ellipsoid or oblong aeciospores

Fig. 69. Rust, caused by *Uromyces appendiculatus*, on the upper leaf surface. (Courtesy M. A. Pastor-Corrales)

Fig. 70. Rust, caused by *Uromyces appendiculatus*, on a pod. (Courtesy H. F. Schwartz)

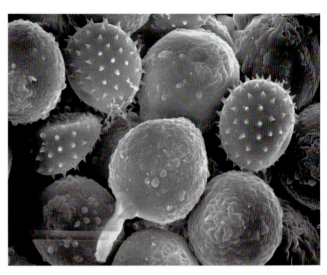

Fig. 71. Rust urediniospores produced by *Uromyces appendiculatus*; scanning electron micrograph. (Courtesy M. A. Pastor-Corrales)

Fig. 72. Pycnia of rust, caused by *Uromyces appendiculatus*, on the upper surface of a leaflet from a naturally infected volunteer bean seedling. (Courtesy H. F. Schwartz)

that are 16–24 × 18–33 µm. Aeciospores infect beans to produce the amphigenous, brown uredinia (Fig. 69). There are few reports of teliospore germination and production of basidiospores, pycnia, and aecia. These portions of the life cycle have only been reported temperate climates.

U. appendiculatus is among the most variable of all plant pathogens. More than 250 races have been identified from around the world and individual field collections of urediniospores often contain several races. A new set of 12 differential cultivars, containing six Andean and six Middle American cultivars, was selected during the 2002 international bean rust workshop to replace the previous set that contained 19. A modified binary system was also chosen for identifying races. The new set of differential cultivars takes into consideration improved information on resistance genes, the existence of Andean and Middle American bean gene pools, and information indicating the existence of Andean-specific pathotypes of *U. appendiculatus*. Coevolution of the rust pathogen and bean host has been suggested in recent studies.

Germination of urediniospores and infection occurs in 10–18 h in the presence of plant surface moisture at optimal temperatures of 16–25°C but does not occur at temperatures above 28°C. The latent period for development of uredinial sporulation ranges from 7 days at the optimum temperature of 24°C to 9 days at 16°C. Prolonged temperatures above 32°C can kill the fungus. Abundant urediniospore production is favored by high humidities below the saturation point, long day lengths, and vigorously growing host tissue. Uredinium size and sporulation efficiency decrease in severe infections. Urediniospores are dispersed by wind currents and can be carried long distances.

Management

The most effective management of bean rust is host resistance. Attainment of stable resistance effective over time is difficult because of the high degree of pathogenic variability in this fungus. Although most bean cultivars are resistant to several races of the pathogen, only a few cultivars are resistant to most local races. Resistant reactions range from immune to necrotic spots to production of uredinia less than 0.3 mm in diameter. Many rust resistance genes are present in common bean, and all identified genes are dominant. Among these, *Ur-3*, *Ur-4*, *Ur-5*, *Ur-6*, *Ur-7*, and *Ur-11* have been used in developing rust-resistant germ plasm and cultivars. Selected races of the rust pathogen have been used to identify specific rust resistance genes and to introgress these genes into beans, resulting in pyramids of multiple genes. Substantial progress in developing beans with multiple rust resistance genes has been realized in the United States.

Rotation with nonhost crops and removal of volunteer beans and infested bean debris reduces initial inoculum levels. Modification of planting dates to reduce exposure of the crop to long dew periods and temperatures favorable to infection can effectively lessen rust severity.

Chlorothalonil and carbamate fungicides are effective when applied before infection occurs or when rust first appears and then at 7- to 10-day intervals until the appropriate time before harvest. Azoxystrobin, boscalid, and pyraclostrobin can reduce rust severity after the appearance of symptoms, but these may not be labeled for use on beans in all countries.

Selected References

Araya, C. M., Alleyne, A. T., Steadman, J. R., Eskridge, K. M., and Coyne, D. P. 2004. Phenotypic and genotypic characterization of *Uromyces appendiculatus* from *Phaseolus vulgaris* in the Americas. Plant Dis. 88:830-836.

Cummins, B. 1978. Rust Fungi on Legumes and Composites in North America. University of Arizona Press, Tucson.

Gold, R. 1983. Activation and pattern of teliospore germination in *Uromyces appendiculatus* var. *appendiculatus* and basidiospore infection of *Phaseolus vulgaris*. Ph.D. thesis. Universitat Konstanz, Konstanz, Federal Republic of Germany.

Gross, P. L., and Venette, J. R. 2001. Overwinter survival of bean rust urediniospores in North Dakota. Plant Dis. 85:226-227.

Groth, J. V., and Mogen, B. D. 1978. Completing the life cycle of *Uromyces phaseoli* var. *typica* on bean plants. Phytopathology 68:1674-1677.

McMillan, M. S., Schwartz, H. F., and Otto, K. L. 2003. Sexual stage development of *Uromyces appendiculatus* and its potential use for disease resistance screening of *Phaseolus vulgaris*. Plant Dis. 87:1133-1138.

Park, S. O., Coyne, D. P., Steadman, J. R., Crosby, K. M., and Brick, M. A. 2004. RAPD and SCAR markers linked to the Ur-6 Andean gene controlling specific rust resistance in common bean. Crop Sci. 44:1799-1807.

Stavely, J. R. 1984. Pathogenic specialization in *Uromyces phaseoli* in the United States and rust resistance in beans. Plant Dis. 68:95-99.

Steadman, J. R., Pastor-Corrales, M. A., and Beaver, J. S. 2002. An overview of the 3rd Bean Rust and 2nd Bean Common Bacterial Blight International Workshops, March 4-8, 2002, Pietermaritzburg, South Africa. Annu. Rep. Bean Improv. Coop. 45:120-124.

(Prepared by J. R. Stavely;
Revised by M. A. Pastor-Corrales and J. R. Steadman)

Scab

Scab occurs in Mexico, Central America, the West Indies, and South Carolina (United States) and is important on common bean in eastern African countries, where yield losses have reached 70%.

Symptoms

Scab can affect all aerial parts of plants, including most types of beans (common, lima, scarlet runner, and mung) and cowpea. Infected leaves, stems, or pods exhibit raised, wartlike protuberances up to 10 mm in diameter that are tan to red or brown. Leaf spots occur along the veins on either leaf surface, turn yellow, and develop slightly raised margins. Stem lesions are brown to gray with yellow to black borders. Pod lesions are brown to purple-black, circular, punctate, and 5 mm in diameter. Pods may be malformed. Conidia form in dark pycnidia in the lesions.

Causal Organism

Scab is caused by the ascomycete *Elsinoe phaseoli* Jenk. The anamorph (conidial state) of *E. phaseoli* is a *Sphaceloma* sp. The hyaline to pale conidia of *E. phaseoli* are 4 × 10 µm. The mycelium is hyaline and submerged; it supports short, stubby

Fig. 73. Aecia of rust, caused by *Uromyces appendiculatus*, on the lower surface of a leaflet from a naturally infected volunteer bean seedling. (Courtesy H. F. Schwartz)

conidiophores that gradually darken and form compact sporodochia. Wrinkled, yellowish to reddish purple or reddish brown colonies form slowly at 22–26°C on acidified potato dextrose agar or 1% cornmeal agar. Ascomata are 30–40 μm long and ascospores are 5–6 × 13–15 μm. Conidia are spherical to elliptical and 1–3 × 3–8 μm. Ascomata may also form on the leaf surface and cover the lesions as dark, punctate bodies 0.1–0.6 mm in diameter. Asci are subglobose to ellipsoid, measure 15–22 × 20–32 μm, and contain ascospores with one to three septa.

Disease Cycle and Epidemiology

The pathogen survives in host debris and within and on the seed coat. Little is known about environmental conditions that favor infection and spread.

Management

Rotation with nonhost crops, such as cereals and corn, may reduce disease severity by reducing initial inoculum levels. Pathogen-free seeds should be planted. Resistance has been identified in common bean. Treatment with chemical seed dressings may reduce transmission of the fungus in contaminated seed lots.

Selected References

Allen, D. J., and Lenné, J. M. 1998. Legume scab fungi. Pages 29-31 in: The Pathology of Food and Pasture Legumes. CAB International, Wallingford, U.K.

Phillips, A. J. L. 1995. Evaluation of some cultivars of *Phaseolus vulgaris* in the greenhouse and in the field for resistance to *Elsinoe phaseoli*. Afr. Plant Prot. 1:59-62.

Phillips, A. J. L. 1996. Variation in pathogenicity among isolates of *Elsinoe phaseoli* from *Phaseolus* species. Ann. Appl. Biol. 128:209-218.

Sivanesan, A., and Holliday, P. 1971. *Elsinoe phaseoli*. Descriptions of Pathogenic Fungi and Bacteria, No. 314. Commonwealth Mycological Institute and Association of Applied Biologists, Kew, Surrey, England.

(Prepared by H. F. Schwartz)

Soybean Rust

Soybean rust, unlike other rusts, attacks many leguminous hosts, including common, scarlet runner, lima, and other kinds of beans. It is especially destructive in soybean, but it is a relatively rare and minor problem in common bean. The disease has been reported in China, Taiwan, Japan, Thailand, India, Australia, Nigeria, South Africa, Mozambique, Zimbabwe, and the United States (Hawaii) in the Eastern Hemisphere and in Brazil, Colombia, Costa Rica, and Puerto Rico in the Western Hemisphere. It was not reported from the continental United States until its introduction via hurricane Ivan in September 2004.

Symptoms

Chlorotic leaf spots develop into angular, tan to reddish brown or purple leaf lesions, 0.2–4.0 mm in diameter, within a week after infection (Fig. 74). Up to 20 tan to brown uredinia, each less than 0.3 mm in diameter, develop in each lesion (Fig. 75). Uredinia open by a pore to produce many pale brown to light tan or nearly white urediniospores. Sporulation occurs predominantly on the abaxial (lower) leaf surface (Fig. 76), especially in the lower to mid levels of the plant canopy. The angular lesions resemble those of common bacterial blight and angular leaf spot, but the soybean rust fungus can be distinguished by microscopic identification of the conical uredinia and characteristic urediniospores.

Causal Organisms

Soybean rust is caused by two species of basidiomycete fungi: *Phakopsora pachyrhizi* Syd., which occurs on 34 natural hosts and 61 experimental hosts in Australia, Asia, and the United States (Hawaii), and *P. meibomiae* (Arth.) Arth., which occurs on 41 natural hosts and 25 experimental hosts in the Western Hemisphere. *P. pachyrhizi* is more aggressive than is *P. meibomiae*. *P. pachyrhizi* is often referred to as Asian soybean rust, and *P. meibomiae* is referred to as American soybean rust. The latter rust has been observed in Central and South America and the Caribbean since the 1970s. Pathotypes of Asian soybean

Fig. 75. Uredinia of the Asian soybean rust fungus, *Phakopsora pachyrhizi*, in one lesion. (Courtesy A. J. Liebenberg and M. M. Liebenberg)

Fig. 76. Sporulation of the Asian soybean rust fungus, *Phakopsora pachyrhizi*, on the lower surface of a bean leaf. (Courtesy A. J. Liebenberg and M. M. Liebenberg)

Fig. 74. Asian soybean rust lesions, caused by *Phakopsora pachyrhizi*, on bean leaves. (Courtesy A. J. Liebenberg and M. M. Liebenberg)

rust have been especially damaging to soybean during recent years.

Soybean is the most important host, and lima bean may be a more common host than is common bean in the Americas. There are many other leguminous hosts, including other *Glycine* spp., *Cajanus cajan* (L.) Millsp., *Canavalia maritima* Thouars, *Canavalia gladiata* (Jacq.) DC., *Lablab purpureus* (L.) Sweet, *Lespedeza juncea* (L.f.) Pers., *Lupinus angustifolius* L., *Macroptilium atropurpureum* (Moc. & Sessé ex DC.) Urb., *Macroptilium lathyroides* (L.) Urb., *Phaseolus coccineus* L., *Vigna unguiculata* (L.) Walp., and *Vigna mungo* (L.) Hepper.

The uredinia of the two species each have a basal peridium terminating in paraphyses. *P. pachyrhizi* is a hemicyclic rust fungus for which pycnial and aecial stages are unknown. The urediniospores are obovoid or ellipsoid, finely and densely echinulate, colorless to pale yellowish brown, and 13–22 × 18–30 µm. Telia develop among the uredinia. They are reddish brown, covered by the host epidermis, and crustose. The telia contain two to five layers of oblong, pale yellow to chestnut brown teliospores that are 5–13 × 14–28 µm.

Urediniospores of *P. meibomiae* are ellipsoid, densely echinulate, colorless to pale yellowish brown, and 12–24 × 16–31 µm. Telia contain teliospores arranged in one to four layers. They are cinnamon brown to light chestnut brown and 1.5–2.0 µm thick.

Disease Cycle and Epidemiology

Urediniospores, the primary means of disease spread, are distributed in the air by wind and rain and can remain viable for 1–2 months, depending upon environmental conditions. A dew period (free moisture) of at least 6 h is required for the urediniospores to germinate and for infection to occur. Urediniospores do not germinate below 8°C or above 30°C. Maximum germination and infection occur at about 20°C. Upon germination, penetration of leaves is direct or through stomates to produce the angular lesion that usually contains multiple uredinia. Production of urediniospores starts about 10 days after infection and continues for several weeks. The optimal temperature for postinfection disease development is about 24°C. During the rainy season in the tropics, the prevalence of *P. pachyrhizi* on beans often increases, whereas that of common bean rust, caused by *Uromyces appendiculatus* (Pers.:Pers.) Unger, may decline. Germination of teliospores has not been reported, and their role in the life cycle is unknown.

Management

Little or no attention has been given to developing management tools for soybean rust in beans because of the unknown importance of the disease in common bean. Certain fungicides are likely to be effective. Many cultivars are susceptible, but if it were screened for, resistance will probably be found in common bean.

Selected References

Anahosur, K. H., and Waller, J. M. 1978. *Phakopsora pachyrhizi*. Descriptions of Pathogenic Fungi and Bacteria, No. 589. Commonwealth Mycological Institute and Association of Applied Biologists, Kew, Surrey, England.

Bromfield, K. R. 1984. Soybean Rust. Monogr. 11. American Phytopathological Society, St. Paul, MN.

du Preez, E. D., van Rij, N. C., Lawrance, K. F., Miles, M. R., and Frederick, R. D. 2005. First report of soybean rust caused by *Phakopsora pachyrhizi* on dry beans in South Aftrica. Plant Dis. 89:206.

Frederick, R. D., Snyder, C. L., Peterson, G. L., and Bonde, M. R. 2002. Polymerase chain reaction assays for the detection and discrimination of the soybean rust pathogens *Phakopsora pachyrhizi* and *P. meibomiae*. Phytopathology 92:217-227.

Hartman, G. L., Sinclair, J. B., and Rupe, J. C., eds. 1999. Compendium of Soybean Diseases, 4th ed. American Phytopathological Society, St. Paul, MN.

Vakili, N. G., and Bromfield, K. R. 1976. Phakopsora rust on soybean and other legumes in Puerto Rico. Plant Dis. Rep. 60:995-999.

(Prepared by J. R. Stavely; Revised by M. A. Pastor-Corrales)

Web Blight

Web blight is prevalent in humid, highland and lowland, tropical and subtropical regions of Latin America where high moisture levels prevail. Yield losses can reach 100%. Seed discoloration often reduces seed quality.

Symptoms

The disease can affect all aerial parts of beans. Infection may originate from basidiospores or from sclerotia or quiescent mycelia from soil debris. Those developing from sclerotia appear as small necrotic spots (5–10 mm in diameter) with brown centers and olive green margins (Fig. 77). These spots become water-soaked, enlarge, and coalesce rapidly to give the infected plant the appearance of having been scalded by hot water (Fig. 78). The surface of infected plant parts, such as leaves, petioles, flowers, and pods, becomes rapidly covered by whitish to brown mycelium, at times accompanied by small sclerotia (Fig. 79). Leaves die and abscise within 3–6 days after infection.

Infections caused by basidiospores appear as distinct, necrotic, circular lesions 2–3 mm in diameter. They are light brown or brick red with a lighter center. Under dry conditions, these round spots fall from the leaf, resulting in a symptom known as cock-eye. Under humid conditions, the round spots give rise to secondary lesions that may expand several millimeters. In comparison to infections originating from sclerotia, these lesions seldom cause defoliation.

Symptoms on pods are similar to those produced by mycelia on foliage (Fig. 80). If infection occurs during pod development or filling, symptoms appear as light brown, irregular lesions that frequently coalesce and cause poorly formed or decayed seeds (Fig. 81). Severe infections can kill the pods. Seeds can be infected by mycelium in the endosperm and radicle end of the embryo with no evident symptoms; however, in severe cases, mycelium can infect the whole seed and blemish or bleach the seed coat surface.

Causal Organism

Web blight is caused by *Rhizoctonia solani* Kühn, the sclerotial or asexual state (anamorph) of the basidiomycete fungus *Thanatephorus cucumeris* (A. B. Frank) Donk. A root rot is also caused by *R. solani* and is discussed within this compendium. The asexual state is differentiated on the basis of hyphal fusion,

Fig. 77. Web blight, caused by *Thanatephorus cucumeris* (anamorph *Rhizoctonia solani*), on leaves. (Courtesy H. F. Schwartz, from the files of M. A. Pastor-Corrales)

known as anastomosis groups (AG). Web blight is caused by subgroups of AG-1 (i.e., AG-1-IE, AG-1-IF, and AG-1-IA) and AG-2 (i.e., AG-2-2 IIIB and AG-2-2 IV). The latter group only infects by basidiospores. *R. solani* hyphae are wide, with multinucleate cells at advancing edges of the colony. Branches arise near the distal end of a cell and develop at approximately a right angle to the subtending hyphae. They are constricted at the point of origin and septate shortly above. Cells are 5–12 × 250 μm and have conspicuous dolipore septa. Some hyphae are differentiated into swollen monilioid cells, often 30 μm or more

in width. Colonies of AG-1 isolates produce variable (0.2–20 mm in diameter), superficial, white sclerotia that turn brown to dark brown and become firm, rough, and subglobose at maturity. Colonies of AG-2-2 IV produce soft sclerotia with indefinite shape and irregular surface.

Whitish basidial fructifications bear a hymenium of discontinuous, barrel-shaped, subcylindrical basidia that are 10–25 × 16–19 μm. Each basidium bears stout, slightly divergent sterigmata (usually four) that may have adventitious septa and are 5.5 × 36.5 μm. Hyaline basidiospores form on the sterigmata and are oblong to broadly ellipsoid and unilaterally flattened, prominently apiculate, smooth, thin walled, and 4–8 × 6–14 μm. Hymenia bearing basidiospores are observed on the abaxial leaf but not on culture media. Basidiospores germinate by repetition (production of secondary basidiospores) or by germ tube.

Disease Cycle and Epidemiology

Mycelia that originate from either sclerotia or basidiospores can initiate web blight. Contaminated seeds and rain-splashed soil particles bearing quiescent mycelium may also serve as inoculum sources.

Penetration of host tissue occurs directly or through stomata by an infection cushion at the hyphal end. Subepidermal hyphae develop intercellularly and intracellularly. Initial symptoms may appear on primary leaves, and hyphal strands may extend to trifoliolate leaves and other plant parts. Sclerotia, cobweb mycelia, or both then form on infected organs and can contrib-

Fig. 78. Coalesced lesions of web blight, caused by *Thanatephorus cucumeris* (anamorph *Rhizoctonia solani*), on a leaf. (Courtesy H. F. Schwartz, from the files of M. A. Pastor-Corrales)

Fig. 80. Web blight, caused by *Thanatephorus cucumeris* (anamorph *Rhizoctonia solani*), on pods and seeds. (Courtesy G. Godoy-Lutz)

Fig. 79. Mycelium and sclerotia of *Thanatephorus cucumeris* (anamorph *Rhizoctonia solani*), cause of web blight, on a stem. (Courtesy H. F. Schwartz, from the files of M. A. Pastor-Corrales)

Fig. 81. Healthy (right) and web blight-affected (left) seeds. Web blight, caused by *Thanatephorus cucumeris* (anamorph *Rhizoctonia solani*). (Courtesy G. Godoy-Lutz)

ute to subsequent cycles of infection and spread to other plant organs or plants, especially during periods of splashing rain. Infections by basidiospores have lower rates of development.

Web blight epidemics are favored by rainy weather, moderate (20°C) to high (30°C) air temperatures, moderate to high soil temperatures, and high relative humidity (greater than 80%).

Management

Certain cultivars are less susceptible to disease than are others, but immunity to infection is not known in common bean. Certified or disease-free seeds can prevent epidemics and long-distance spread. Cultural practices, such as rotation with non-host crops, may reduce soil inoculum levels. Use of mulches may reduce the levels of initial inoculum, splashing of sclerotia, or infested soil debris, but it is not effective where basidiospores are the main source of inoculum. Mulches can also promote slug damage. Changing planting dates under irrigated conditions can reduce disease incidence. Chemical sprays with benzimidazoles, carbendazim, and captafol can reduce disease severity and yield losses.

Selected References

Allen, D. J. 1983. The Pathology of Tropical Food Legumes: Disease Resistance in Crop Improvement. J. Wiley & Sons, New York.
Galvez, G. E., Mora, B., and Pastor-Corrales, M. A. 1989. Web blight. Pages 195-209 in: Bean Production Problems in the Tropics, 2nd ed. H. F. Schwartz and M. A. Pastor-Corrales, eds. Centro Internacional de Agricultura Tropical (CIAT), Cali, Colombia.
Godoy-Lutz, G., Arias, J., Steadman J. R., and Eskridge, K. M. 1996. Role of natural seed infection by the web blight pathogen in common bean seed damage, seedling emergence, and early disease development. Plant Dis. 80:887-890.
Godoy-Lutz, G., Steadman, J. R., Higgins, B., and Powers, K. 2003. Genetic variation among isolates of the web blight pathogen of common bean based on PCR-RFLP of the ITS-rDNA region. Plant Dis. 87:766-771.
Holliday, P. 1980. Fungus Diseases of Tropical Crops. Cambridge University Press, New York.
Kuninaga, S., Natsuaki, T., Takeuchi, T., and Yokosawa, R. 1997. Sequence variation of the rDNA ITS regions within and between anastomosis groups in *Rhizoctonia solani*. Curr. Genet. 32:237-243.
Sneh, B., Jabaji-Hare, S., Neate, S., and Dijst, G., eds. 1996. *Rhizoctonia* Species: Taxonomy, Molecular Biology, Ecology, Pathology and Disease Control. Kluwer Academic Publishers, Dordrecht, the Netherlands.

(Prepared by H. F. Schwartz;
Revised by G. Godoy-Lutz and J. R. Steadman)

White Leaf Spot

The disease is present in Latin American countries, such as Colombia, Guatemala, and the Dominican Republic, and in North America in the United States (Minnesota). Yield losses have exceeded 40% in Colombia.

Symptoms

Symptoms appear first on the abaxial (lower) leaf surface as white, angular spots (2–5 mm across) restricted by veins. Angular, white, velvety spots bearing conidiophores and conidia may also occur on the adaxial (upper) leaf surface (Fig. 82) and eventually enlarge and coalesce. Spots become slightly gray with age and are pale green to chlorotic on the opposing leaf surface. Leaf necrosis and defoliation may occur (Fig. 83).

Causal Organism

White leaf spot is caused by *Pseudocercosporella albida* (Matta & Belliard) Deighton (anamorph *Cercospora albida* Matta & Belliard). Conidiophores emerge through stomata in dense fascicles of 30–200 each. They lack stroma or are slightly stromatic; are thick, short, unbranched, and subhyaline; rarely have one or two septa; and are irregularly swollen, slightly geniculate, subtruncate, and 4–6 × 20–60 μm. Conidia are hyaline, narrow walled, cylindrical-obclavate, slightly and irregularly curved, subtruncate at the base, and rounded at the apex. Conidia have zero to five septa and are 2–4 × 30–125 μm.

Disease Cycle and Epidemiology

Leaves, especially the lower leaves, become infected during periods of low to moderate moisture levels and low temperatures, particularly at altitudes higher than 1,500 m above sea level (masl) in Colombia. The fungus apparently survives in previously infected crop debris.

Management

Certain bean cultivars and lines, such as Mexico 6 and PI 313755, are resistant. Rotation with nonhost crops, such as cereals and corn, may reduce disease severity by reducing initial inoculum levels. Sprays with fungicides may be effective.

Selected References

del Río, L. E., Bradley, C. A., and Lamppa, R. S. 2003. First report of white leaf spot of dry bean caused by *Pseudocercosporella albida* in North America. Plant Dis. 87:1537.
Schwartz, H. F., Correa V., F., Pineda D., P. A., Otoya, M. M., and Katherman, M. J. 1981. Dry bean yield losses caused by Ascochyta, angular, and white leaf spots in Colombia. Plant Dis. 65:494-496.

(Prepared by H. F. Schwartz)

Fig. 82. White leaf spot, caused by *Pseudocercosporella albida*. (Courtesy H. F. Schwartz)

Fig. 83. White leaf spot, caused by *Pseudocercosporella albida*. (Courtesy H. F. Schwartz)

White Mold

White mold occurs in most areas of the world where beans are grown, except the warm, humid tropics, and is often highly destructive. Crop losses may reach 100%. The disease typically becomes serious in crops that have a dense canopy, in fields with a history of the disease, and during seasons when cool, moist conditions occur during and after flowering.

Symptoms

Infected flowers may develop a white, cottony appearance as mycelium grows on the surface. Lesions on pods, leaves, branches, and stems are initially small, circular, dark green, and water-soaked but rapidly increase in size, may become slimy, and may eventually encompass and kill the entire organ (Fig. 84). Under moist conditions, these lesions may also develop a white, cottony growth of external mycelium (Fig. 85). Affected tissues dry out and bleach to a beige or white that contrasts with the normal light tan color of senescent tissue. The epidermis easily sloughs off when the stem or pod is rubbed. Cushions of white mycelium (immature sclerotia) develop into hard, black sclerotia in and on infected tissue (Fig. 86). Entire branches or plants may be killed.

Causal Organism

White mold is caused by *Sclerotinia sclerotiorum* (Lib.) de Bary, an ascomycete fungus that infects more than 400 plant species. Mycelium is hyaline, septate, and branched. Sclerotia (Fig. 87) are globose to C-shaped to cylindrical ($2-15 \times 2-30$ mm), depending on where they are formed in or on the plant, with a black outer rind and a white inner cortex. One or more apothecia may arise from a sclerotium (Fig. 88). Apothecia are ocher to light tan in color. The receptacle is 2–10 mm wide, flat to concave when young, and flat to convex at maturity; it tapers to form a stipe 1–2 mm wide and 3–30 mm long. Asci are cylindrical, have a thickened apex possessing a pore channel, and contain eight ascospores. Ascospores are $4-6 \times 9-14$ μm, uniseriate, hyaline, ellipsoid, and biguttulate and contain two to four nuclei. Eight to ten million ascospores may be produced from an apothecium. Microconidia are globose, hyaline, and 2–4 μm in diameter and are produced from phialides in sporodochia, on hyphae, or from the surface of the hymenium or culture. The role of microconidia is not known and conidia are not produced. *S. sclerotiorum* is homothallic and exhibits clonality.

Disease Cycle and Epidemiology

Sclerotia may survive in soil for 5 or more years. Under suitable conditions of temperature, light, and moisture, sclerotia within 5 cm of the soil surface germinate to produce a stipe or stipes that develops an apothecium or apothecia, respectively.

Fig. 84. White mold on a stem and branches infected with *Sclerotinia sclerotiorum*. (Courtesy R. L. Forster)

Fig. 85. White mold and sclerotia of *Sclerotinia sclerotiorum* on pods. (Courtesy H. F. Schwartz)

Fig. 86. Sclerotia of *Sclerotinia sclerotiorum*. (Courtesy R. L. Forster)

Fig. 87. *Sclerotinia sclerotiorum* mycelium and sclerotia on potato dextrose agar. (Courtesy H. F. Schwartz)

Ascospores are released from turgid asci, often simultaneously, by "puffing". Ascospores germinate and colonize flowers and other tissues that are senescing, and the mycelium from colonized tissues invades adjacent organs. Flower parts often fall onto pods, leaves, branches, and stems and provide nutrients required by the fungus to penetrate these organs. Infected tissues are rapidly killed and become dry and bleached. Limited spread of the fungus from one plant to another may occur by mycelial growth between adjoining tissues. Sclerotia form in or on infected tissues and may fall to the soil, remain in crop debris, or be removed with harvested seeds or pods. The fungus may continue to grow and cause disease in green bean in transit and in storage (nesting).

Sclerotia are preconditioned to produce apothecia by exposure to moist, cool (4°C), or freezing conditions for several weeks. To produce apothecia, preconditioned sclerotia require moist soil (water potentials greater than –5 bars) for one to several weeks at temperatures of 11–20°C. Apothecia can produce ascospores for 5–10 days. Apothecia generally do not develop or persist until a dense crop canopy has formed to provide a cool, moist microclimate. Ascospores may be released by changes in relative humidity or by physical disturbance of the apothecium. Most ascospores produced within a field are deposited within the crop canopy. Externally produced windblown ascospores may also contribute to limited infection in the bean crop.

Dissemination of the pathogen occurs by aerial transport of ascospores; irrigation water carrying mycelium, ascospores, and sclerotia; and infected or infested seeds (Fig. 89). The wide host range of the pathogen leads to widespread contamination of fields through crop or weed infection and to the subsequent production of sclerotia.

Ascospores can survive on plant surfaces for up to 2 weeks, and mycelium in infected blossoms may remain viable for 1 month. Disease develops at temperatures from 5 to 30°C and most rapidly at 20–25°C, especially in the presence of moisture.

Management

Some bean cultivars and breeding lines, such as Ex Rico 23 and G122, are less susceptible than are other cultivars, but immunity to white mold is not known in common bean. Resistance has been found in *Phaseolus coccineus* L. and is being transferred to *P. vulgaris* L. Plants with an upright growth habit and an open architecture can avoid the disease.

Rotation with nonhost crops, such as cereals and corn, may reduce disease severity by reducing the prevalence of initial inoculum. However, inoculum may be present from a previous host; maintained on weeds; or carried into the field by wind, irrigation water, or insects, such as bees. The disease severity may be reduced by practices that reduce plant surface and soil moisture levels and reduce canopy density. Examples of these practices are planting rows parallel to the prevailing wind, avoiding excessive and late-season irrigation and excessive amounts of nitrogenous fertilizer, and using wide row spacing. Timely harvest, followed by rapid cooling and storage of healthy pods at 7–10°C, can provide simple and effective management of white mold in snap beans.

Coniothyrium minitans Campb. and other biological control agents break down or destroy overwintering sclerotia in the soil and may provide possible future disease management strategies.

Fungicide sprays, such as thiophanate methyl and boscalid, applied during the flowering period is another disease management strategy. However, disease management is difficult with dry edible beans where plants develop lush, viny growth and the flowering period extends over several weeks. Sprays applied before or after flowering are less effective.

Selected References

Boland, G. J., and Hall, R. 1987. Epidemiology of white mold of white bean in Ontario. Can. J. Plant Pathol. 9:218-224.

Boland, G. J., and Hall, R. 1994. Index of plant hosts of *Sclerotinia sclerotiorum*. Can. J. Plant Pathol. 16:93-108.

British Society for Plant Pathology. 2001. The 11th International Sclerotinia Workshop. BSPP News 40(Autumn). Online abstracts (www.bspp.org.uk). The Society, Reading, Berks, U.K.

del Río, L. E., Venette, J. R., and Lamey, H. A. 2004. Impact of white mold incidence on dry bean yield under nonirrigated conditions. Plant Dis. 88:1352-1356.

Kohli, Y., Morrall, R. A. A., Anderson, J. B., and Kohn, L. M. 1992. Local and trans-Canadian clonal distribution of *Sclerotinia sclerotiorum* on canola. Phytopathology 82:875-880.

Kohli, Y., Brunner, L. J., Yoell, H., Milgroom, M. G., Anderson, J. B., Morrall, R. A. A., and Kohn, L. M. 1995. Clonal dispersal and spatial mixing in populations of the plant pathogenic fungus *Sclerotinia sclerotiorum*. Mol. Ecol. 4:69-77.

Kolkman, J. M., and Kelly, J. D. 2003. QTL conferring resistance and avoidance to white mold (*Sclerotinia sclerotiorum*) in common bean (*Phaseolus vulgaris*). Crop Sci. 43:539-548.

Miklas, P. N., Delorme, R., and Riley, R. 2003. Identification of QTL conditioning resistance to white mold in a snap bean population. J. Am. Soc. Hortic. Sci. 128:564-570.

Morton, J. G., and Hall, R. 1989. Factors determining the efficacy of chemical control of white mold in white bean. Can. J. Plant Pathol. 11:297-302.

Steadman, J. R. 1983. White mold—A serious yield-limiting disease of bean. Plant Dis. 67:346-350.

Tu, J. C. 1997. An integrated control of white mold (*Sclerotinia sclerotiorum*) of beans with emphasis on recent advances in biological control. Bot. Bull. Acad. Sin. 38:73-76.

Willetts, H. J., and Wong, J. A.-L. 1980. The biology of *Sclerotinia*

Fig. 88. Sclerotium and apothecia of *Sclerotinia sclerotiorum*. (Courtesy R. Hall, from the files of G. J. Boland)

Fig. 89. Sclerotia of and seeds infected with *Sclerotinia sclerotiorum*. (Courtesy R. Hall)

sclerotiorum, S. trifoliorum, and *S. minor* with emphasis on specific nomenclature. Bot. Rev. 46:101-165.

(Prepared by R. Hall and J. R. Steadman;
Revised by J. R. Steadman and G. Boland)

Yeast Spot

Yeast spot or seed pitting can be a seed production problem in the West Indies, in the United States, and in Latin American countries, such as Brazil, Costa Rica, Ecuador, and Peru. Yield losses can reach 100%, depending upon its effect on seed quality and commercial appeal for beans such as lima bean.

Symptoms

Symptoms appear after insects, such as stink bugs (*Megalotomus parvus* Westwood) and lygus bugs (*Lygus hesperus* Knight and *L. elisus* Van Duzee), transmit the causal organism to pods and developing seeds. During feeding, the insects puncture the developing seeds and transfer fungal propagules to the wound sites. The spores germinate and infect the seeds, producing irregular, slightly sunken lesions about 1 mm in diameter. The lesions may be rose, tan, or brown. There may be additional damage to seeds from insect toxins.

Causal Organism

Yeast spot is caused by *Nematospora coryli* Peglion, *N. gossypii* S. Ashby & W. Nowell, and *Eremothecium cymbalariae* Borzí. *N. coryli* develops elliptical cells 6–10 μm wide and 8–14 μm long, followed by mature spherical cells of 20 μm in diameter and myceliumlike strands that measure 2.5–3.5 × 90–140 μm. Culture growth is favored at 25–30°C.

Disease Cycle and Epidemiology

The disease is favored by the presence of weed hosts for the insects.

Management

Management measures consist of eliminating weed hosts, controlling insect populations, and planting clean seeds.

Selected References

Schwartz, H. F. 1989. Yeast spot. Pages 247-248 in: Bean Production Problems in the Tropics, 2nd ed. H. F. Schwartz and M. A. Pastor-Corrales, eds. Centro Internacional de Agricultura Tropical (CIAT), Cali, Colombia.

Zaumeyer, W. J., and Thomas, H. R. 1957. A monographic study of bean diseases and methods for their control. U.S. Dep. Agric. Tech. Bull. 868.

(Prepared by H. F. Schwartz)

Diseases Caused by Bacteria

There are three major bacterial diseases of common bean: bacterial brown spot, common bacterial blight (and its variant, fuscous blight), and halo blight. While few in number, these diseases have had a major impact on bean production throughout the world, and considerable efforts have been expended to manage them, especially through pathogen-free seed programs. Recently, some progress has been made in developing beans with resistance to these diseases. Two other bacterial diseases of lesser or more sporadic impact are bacterial wilt and wildfire.

Bacterial Brown Spot

Bacterial brown spot (or brown spot) is reported from the United States and Brazil and can reach damaging levels in Wisconsin on both snap and dry beans.

Symptoms

Leaf lesions are generally small (approximately 1 mm in diameter), circular, brown, and necrotic and are often surrounded by a yellow zone (Figs. 90 and 91). Lesions may enlarge, coalesce, and occasionally fall out, giving the leaves a ragged appearance. Water-soaking of the tissue is generally absent or minimal. Stem lesions are occasionally observed when the pathogen develops systemically. Lesions on pods are circular and initially water-soaked; they become brown and necrotic (Fig. 92). Infected pods may be distorted where lesions develop. Occasionally, ring spots of lesions occur around a central lesion.

Causal Organism

Brown spot is caused by *Pseudomonas syringae* pv. *syringae* van Hall, a gram-negative, rod-shaped bacterium that is motile,

Fig. 90. Bacterial brown spot, caused by *Pseudomonas syringae* pv. *syringae*, on upper and lower leaf surfaces. (Courtesy H. F. Schwartz)

Fig. 91. Bacterial brown spot, caused by *Pseudomonas syringae* pv. *syringae*, on leaves. (Courtesy H. F. Schwartz)

with a polar tuft of flagella. The pathogen produces fluorescent pigments, is arginine-dihydrolase negative, and is aerobic. The following compounds are utilized: betaine, *N*-caprate, *N*-caproate, *N*-caprylate, citrate, D-gluconate, glutarate, DL-glycerate, glycerol, DL-β-hydroxybutyrate, *meso*-inositol, iso-ascorbate, L-proline, sorbitol, sucrose, and *meso*-tartrate. Growth is maximal at 28–30°C and appears as cream-colored, translucent colonies. Pathogenic isolates produce a bacteriocin known as syringacin W-l in the host plant. *P. syringae* pv. *syringae* has an extremely wide host range that includes common bean, fava bean, lima bean, pea, soybean, *Lablab purpureus* (L.) Sweet, *Pueraria lobata* (Willd.) Ohwi, *Vigna sesquipedalis* (L.) Fruw., and *V. unguiculata* (L.) Walp. However, only isolates from beans are highly pathogenic to beans.

Disease Cycle and Epidemiology

P. syringae pv. *syringae* can multiply and survive on a number of crop (e.g., lima bean and pea) and noncrop (e.g., *Vicia villosa* Roth) species, which then serve as sources of primary inoculum for infection of beans. It is spread during rainstorms, and its severity is directly related to the intensity of water droplets impacting the leaf. Sprinkler irrigation is also an effective means for spread of the pathogen. The bacterium can also contaminate seeds, but the role of such inoculum in disease development is not well understood.

Management

Weeds, volunteer beans, and other plants that may serve as reservoirs for *P. syringae* pv. *syringae* should be managed. Pathogen-free seeds and resistant cultivars (e.g., great northern Weihing) should be planted. Bactericidal sprays containing fixed copper should be applied to manage secondary spread and epiphytic growth of *P. syringae* pv. *syringae* during the late-vegetative to early-flowering periods of plant development.

Selected References

Daub, M. E., and Hagedorn, D. J. 1981. Epiphytic populations of *Pseudomonas syringae* on susceptible and resistant bean lines. Phytopathology 71:547-550.
Ercolani, G. L., Hagedorn, D. J., Kelman, A., and Rand, R. E. 1974. Epiphytic survival of *Pseudomonas syringae* on hairy vetch in relation to epidemiology of bacterial brown spot of bean in Wisconsin. Phytopathology 64:1330-1339.
Hirano, S. S., Rouse, D. I., Clayton, M. K., and Upper, C. D. 1995. *Pseudomonas syringae* pv. *syringae* and bacterial brown spot of snap bean: A study of epiphytic phytopathogenic bacteria and associated disease. Plant Dis. 79:1085-1093.
Lindemann, J., Arny, D. C., and Upper, C. D. 1984. Epiphytic populations of *Pseudomonas syringae* pv. *syringae* on snap bean and nonhost plants and the incidence of bacterial brown spot disease in relation to cropping patterns. Phytopathology 74:1329-1333.

Fig. 92. Bacterial brown spot, caused by *Pseudomonas syringae* pv. *syringae*, on pods. (Courtesy H. F. Schwartz)

Mohan, S. K., and Hagedorn, D. J. 1989. Additional bacterial diseases. Pages 303-319 in: Bean Production Problems in the Tropics, 2nd ed. H. F. Schwartz and M. A. Pastor-Corrales, eds. Centro Internacional de Agricultura Tropical (CIAT), Cali, Colombia.

(Prepared by A. W. Saettler;
Revised by C. Ishimaru, S. K. Mohan, and G. D. Franc)

Common Bacterial Blight

Common bacterial blight (or common blight) affects the foliage and pods of beans and is considered to be a major problem in most production areas of the world. During extended periods of warm, humid weather, the disease can be highly destructive, causing losses in both yield and seed quality. Common blight typically develops when contaminated seeds are planted, when plantings are made in fields with a history of the disease, and when the climate is consistently hot and wet or humid.

Symptoms

Leaf symptoms initially appear as water-soaked spots (Fig. 93) that gradually enlarge, become flaccid, and then necrotic and are often bordered by a small zone of lemon yellow tissue (Fig. 94). Lesions can be found along margins and in interveinal areas of the leaf. As lesions enlarge and coalesce, the plants appear to be burned. In severe infections, dead leaves remain attached to the plants at maturity. Bacteria from infested material provide inoculum for secondary spread. Pod symptoms consist of lesions that are generally circular, slightly

Fig. 93. Water-soaked lesion of common bacterial blight, caused by *Xanthomonas axonopodis* pv. *phaseoli*. (Courtesy H. F. Schwartz)

Fig. 94. Common bacterial blight, caused by *Xanthomonas axonopodis* pv. *phaseoli*, on upper (left) and lower (right) leaf surfaces. (Courtesy H. F. Schwartz)

sunken, and dark red-brown (Fig. 95). Lesions vary in size and shape depending on pod age. Under highly humid conditions, pod lesions are frequently covered with bacterial ooze. Symptoms on white seeds are evident as butter yellow or brown spots distributed throughout the seed coat or restricted to the hilum area (Fig. 96). Severely affected seeds are frequently shriveled and exhibit poor germination and vigor.

Causal Organism

Common blight is caused by *Xanthomonas campestris* pv. *phaseoli* (Smith) Dye (syn. *Xanthomonas axonopodis* pv. *phaseoli* (Smith) Vauterin et al.), which includes its fuscous variant that produces a brown pigment in culture. *X. campestris* pv. *phaseoli* is a gram-negative, straight rod that is aerobic and motile by a polar flagellum. It produces acid from arabinose, cellobiose, galactose, glucose, mannose, and trehalose. It also produces H_2S, causes proteolysis of milk, and hydrolyzes starch. *X. campestris* pv. *phaseoli* produces a yellow, non-water-soluble carotenoid pigment (xanthomonadin) and mucoid growth on nutrient glucose agar. Several media (e.g., nutrient agar, yeast dextrose carbonate agar, yeast extract peptone glucose agar amended with antibiotics, and medium for *X. campestris* pv. *phaseoli* [MXP]) are available for rapid isolation and identification of the pathogen.

The pathogen can infect common bean, scarlet runner bean, tepary bean, soybean, *Dolichos lablab* L., *Lupinus polyphyllus* Lindl., *Stizolobium deeringianum* Bort, *Strophostyles helvola* (L.) Elliott, *Vigna aconitifolia* (Jacq.) Marechal, *V. angularis* (Willd.) Ohwi & H. Ohashi, *V. mungo* (L.) Hepper, *V. radiata* (L.) R. Wilcz, and *V. unguiculata* (L.) Walp. under natural or experimental conditions.

Fig. 95. Common bacterial blight, caused by *Xanthomonas axonopodis* pv. *phaseoli*, on pods. (Courtesy S. K. Mohan)

Fig. 96. Common bacterial blight, caused by *Xanthomonas axonopodis* pv. *phaseoli*, on seeds. (Courtesy A. W. Saettler)

Disease Cycle and Epidemiology

Several types of primary inoculum are involved in initiation of common blight. Contaminated seeds constitute a major source of primary inoculum worldwide and are an effective means for both local and long-distance dissemination of the pathogen. *X. campestris* pv. *phaseoli* is an efficient colonizer of both susceptible and tolerant genotypes. Seed contamination by *X. campestris* pv. *phaseoli* is both internal and external. The latter may be managed by applying bactericides (such as streptomycin) to the seeds. Seedlings arising from contaminated seeds support large populations of the pathogen, and *X. campestris* pv. *phaseoli* infects the cotyledons and new leaves as they unfold along the stem axis.

Overwintering of *X. campestris* pv. *phaseoli* in infested plant debris occurs in some temperate production regions. Survival is generally longer in residue at or near the soil surface than it is in residue buried beneath the soil surface. Infested residue is particularly important in the tropics as a source of inoculum from which *X. campestris* pv. *phaseoli* can multiply and survive as an epiphyte on volunteer beans, perennial hosts, and intercropped plants. The bacterium grows epiphytically on leaves of several nonhost crop species and weeds, which serve as additional reservoirs of inoculum. *X. campestris* pv. *phaseoli* is a warm-climate pathogen that causes greatest damage to plants at 28–32°C. High humidity, rain, or both, favor rapid progress of the disease in the field. The time between initial infection and production of inoculum for secondary spread is 10–14 days. *X. campestris* pv. *phaseoli* is spread by windblown rain, soil, and plant debris; contact between wet leaves; irrigation water; people; animals; and insects, such as whiteflies and leaf miners.

Management

Pathogen-free seeds that have been inspected during production in arid areas and tested for freedom from *X. campestris* pv. *phaseoli* should be used; seeds of both resistant and susceptible genotypes should be certified pathogen-free. Seeds should be treated with an antibiotic, such as streptomycin. Crop rotation should be used, with at least 2 years between bean crops. Weeds, volunteer beans, and other alternate hosts of *X. campestris* pv. *phaseoli* should be eliminated through use of herbicides or hand labor. Cultivars with partial resistance (e.g., great northern Harris) should be used where available. Tepary bean (*Phaseolus acutifolius* A. Gray) is a source of immunity to *X. campestris* pv. *phaseoli*. High levels of resistance from multiple sources are available in breeding lines. Foliage should be sprayed with a protectant copper-based bactericide before symptoms appear.

Selected References

Claflin, L. E., Vidaver, A. K., and Sasser, M. 1987. MXP, a semiselective medium for *Xanthomonas campestris* pv. *phaseoli*. Phytopathology 77:730-734.

Gent, D. H., Lang, J. M., and Schwartz, H. F. 2005. Epiphytic survival of *Xanthomonas axonopodis* pv. *allii* and *X. axonopodis* pv. *phaseoli* on leguminous hosts and onion. Plant Dis. 89:558-564.

Miklas, P. N., Coyne, D. P., Grafton, K. F., Mutlu, N., Reiser, J., Lindgren, D., and Singh, S. P. 2003. A major QTL for common bacterial blight resistance derives from the common bean great northern landrace cultivar Montana No. 5. Euphytica 131:137-146.

Saettler, A. W. 1989. Common bacterial blight. Pages 261-283 in: Bean Production Problems in the Tropics, 2nd ed. H. F. Schwartz and M. A. Pastor-Corrales, eds. Centro Internacional de Agricultura Tropical (CIAT), Cali, Colombia.

Schaad, N. W., Jones, J. B., and Chun, W., eds. 2001. Laboratory Guide for Identification of Plant Pathogenic Bacteria, 3rd ed. American Phytopathological Society, St. Paul, MN.

Schuster, M. L., and Coyne, D. P. 1981. Biology, epidemiology, genetics and breeding for resistance to bacterial pathogens of *Phaseolus vulgaris* L. Hortic. Rev. 3:28-58.

Sheppard, J. W., Roth, D. A., and Saettler, A. W. 1989. Detection of *Xanthomonas campestris* pv. *phaseoli* in bean. Pages 17-29 in: Detection of Bacteria in Seed and Other Planting Material. A. W. Saettler, N. W. Schaad, and D. A. Roth, eds. American Phytopathological Society, St. Paul, MN.

Singh, S. P., and Munoz, C. G. 1999. Resistance to common bacterial blight among *Phaseolus* species and common bean improvement. Crop Sci. 39:80-89.

Steadman, J. R., Pastor-Corrales, M. A., and Beaver, J. E. 2002. An overview of the 3rd Bean Rust and 2nd Bean Common Bacterial Blight International Workshops. March 4-8, 2002. Pietermaritzburg, South Africa. Annu. Rep. Bean Improv. Coop. 45:120-124.

Webster, D. M., Atkin, J. D., and Cross, J. E. 1983. Bacterial blights of snap beans and their control. Plant Dis. 67:935-940.

(Prepared by A. W. Saettler;
Revised by C. Ishimaru, S. K. Mohan, and G. D. Franc)

Halo Blight

Halo blight attacks foliage and pods of beans and is a major disease problem worldwide. The disease is most destructive in areas where temperatures are moderate and where abundant inoculum is available.

Symptoms

Leaf symptoms appear several days after infection as small, water-soaked spots on the lower surface. They rapidly become necrotic and are visible on upper and lower leaf surfaces (Fig. 97). Necrotic spots generally remain small (1–2 mm in diameter). However, infection of expanding leaves may result in leaf distortion. A chlorotic zone of yellow-green tissue (halo) may appear around necrotic spots and, in cases of severe infection, plants also may develop a generalized systemic chlorosis (Fig. 98). These chlorotic symptoms are more pronounced at 18–23°C because of temperatures favorable for production of the non-host-specific phaseolotoxin. At temperatures above 23°C, chlorosis associated with phaseolotoxin may become less noticeable or absent. At 7–10 days after infection, bacteria ooze from substomatal cavities to give lesions a greasy, water-soaked appearance. Bacteria are thus available for secondary spread of the disease.

Stems and pods also show evidence of infection. Pod symptoms generally consist of red or brown lesions that may also appear water-soaked. As pods mature and turn yellow to tan, pod lesions may remain green and may exhibit crusty bacterial ooze on the surface (Fig. 99). Seeds may be shriveled or discolored if lesions expand to involve the pod suture (Fig. 100).

Causal Organism

Halo blight is caused by the gram-negative, rod-shaped bacterium *Pseudomonas syringae* pv. *phaseolicola* (Burkholder) Young et al. *P. syringae* pv. *phaseolicola* is strictly aerobic, has a negative reaction for oxidase and arginine dihydrolase, and produces diffusible fluorescent pigments in iron-deficient culture media. The bacterium is able to utilize D-gluconate, L(+)-arabinose, sucrose, succinate, DL-β-hydroxybutyrate, *trans*-aconitate, L-serine, and L-*p*-hydroxybenzoate. Maximal growth and production of phaseolotoxin occur at 20–23°C. Bacterial growth on standard culture media consists of white to cream-colored colonies. Essentially all pathogenic isolates of *P.*

Fig. 98. Extensive yellowing of plant affected with halo blight, caused by *Pseudomonas syringae* pv. *phaseolicola*, on leaves. (Courtesy H. F. Schwartz, from the files of M. A. Pastor-Corrales)

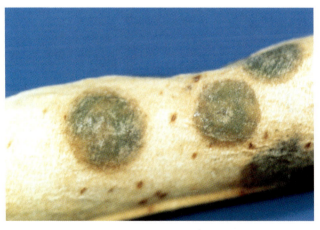

Fig. 99. Lesions and bacterial crust of halo blight, caused by *Pseudomonas syringae* pv. *phaseolicola*, on a pod. (Courtesy A. W. Saettler)

Fig. 100. Halo blight, caused by *Pseudomonas syringae* pv. *phaseolicola*, on a pod and seeds. (Courtesy H. F. Schwartz)

Fig. 97. Halo blight, caused by *Pseudomonas syringae* pv. *phaseolicola*, on leaves. (Courtesy H. F. Schwartz)

syringae pv. *phaseolicola* produce phaseolotoxin. *P. syringae* pv. *phaseolicola* exhibits considerable pathogenic variation in nature, which complicates efforts to identify broad disease resistance. Numerous pathogenic races have been identified based on differential host reactions. Additional hosts infected experimentally include azuki bean, lima bean, mung bean, scarlet runner bean, soybean, tepary bean, *Macroptilium atropurpureum* (Moc. & Sessé ex DC.) Urb., *Phaseolus polyanthus* Greenman, *P. polystachyus* (L.) Britton, Sterns & Poggenb., and *Pueraria lobata* (Willd.) Ohwi.

Disease Cycle and Epidemiology

Sources of *P. syringae* pv. *phaseolicola* inoculum include contaminated seeds and infested plant residue (Fig. 101). As with common blight, seedlings arising from contaminated seeds harbor large numbers of the pathogen before symptom development. Bacteria from infected plant residue are moved to healthy plant tissues by splashing rain or blowing soil particles. The pathogen enters plants through natural openings, such as stomata and hydathodes, or through wounds during periods of high humidity or when free moisture occurs on the plant surface. *P. syringae* pv. *phaseolicola* grows epiphytically on the leaves of both resistant and susceptible genotypes, even when there are no obvious disease symptoms. Secondary spread occurs by splash dispersal during rain and by contact between adjacent leaves wet from irrigation, rain, or dew. Blossoms, pods, and stem tissues are also colonized under favorable conditions. Halo blight is considered to be a cool-temperature disease, and epidemic potential is greatest at 18–22°C. The disease is frequently detected shortly after seedling emergence during conducive weather if inoculum levels are high.

Management

Seeds from plants that have been inspected during production in an arid climate and tested for freedom from *P. syringae* pv. *phaseolicola* should be used. A combined biological enrichment and polymerase chain reaction bioassay has been developed to amplify and detect the phaseolotoxin gene in contaminated seeds. Seeds also may be treated with an antibiotic, such as streptomycin, to reduce surface-borne inoculum. Cultivars with resistance to local races should be planted. Crop rota-

tion with nonhosts of two to several years and the incorporation of bean debris into soil encourage decay of infested debris and the destruction of inoculum. Weeds and volunteer beans also should be eliminated to remove potential reservoirs of *P. syringae* pv. *phaseolicola* inoculum in the field. Overhead irrigation is likely to increase secondary spread compared with furrow irrigation and should be avoided where possible. Cultivation or any other traffic through fields, including inspections by man, should be avoided when foliage is wet to preclude redistribution of inoculum. Properly timed bactericidal spray containing fixed copper may reduce epiphytic populations sufficiently during late-vegetative to early-flowering periods of plant development to reduce disease severity and minimize secondary disease spread.

Selected References

Fourie, D. 1998. Characterization of halo blight races on dry beans in South Africa. Plant Dis. 82:307-310.
Garrett, K. A., and Schwartz, H. F. 1998. Epiphytic *Pseudomonas syringae* on dry beans treated with copper-based bactericides. Plant Dis. 82:30-35.
Katherman, M. J., Wilkinson, R. E., and Beer, S. V. 1980. Resistance and seed infection in three dry bean cultivars exposed to a halo blight epidemic. Plant Dis. 64:857-859.
Schaad, N. W., Cheong, S. S., Tamaki, S., Hatziloukas, E., and Panopoulos, N. J. 1995. A combined biological and enzymatic amplification (BIO-PCR) technique to detect *Pseudomonas syringae* pv. *phaseolicola* in bean seed extracts. Phytopathology 85:243-248.
Schwartz, H. F. 1989. Halo blight. Pages 285-301 in: Bean Production Problems in the Tropics, 2nd ed. H. F. Schwartz and M. A. Pastor-Corrales, eds. Centro Internacional de Agricultura Tropical (CIAT), Cali, Colombia.

(Prepared by A. W. Saettler;
Revised by C. Ishimaru, S. K. Mohan, and G. D. Franc)

Bacterial Wilt

Bacterial wilt of beans has been infrequently reported in the United States. The disease has been reported from Tunisia, Turkey, southern Australia, Belgium, Bulgaria, Greece, Hungary, Romania, the former USSR, the former Yugoslavia, Canada (Alberta and Ontario), Mexico, and Colombia.

Symptoms

In the field, infected plants exhibit wilting or flaccid leaves during periods of moisture stress. Wilting results when patho-

Fig. 101. Bean debris affected with halo blight, caused by *Pseudomonas syringae* pv. *phaseolicola*. (Courtesy H. F. Schwartz)

Fig. 102. Bacterial wilt, caused by *Curtobacterium flaccumfaciens*, on foliage. (Courtesy H. F. Schwartz)

Fig. 103. Water-soaking on stem infected with *Curtobacterium flaccumfaciens*, cause of bacterial wilt. (Courtesy H. F. Schwartz)

Fig. 104. Seed infected with orange (upper right), violet (upper left), and yellow (lower left) variants of *Curtobacterium flaccumfaciens*, cause of bacterial wilt. Healthy seed is on the lower right. (Courtesy A. W. Saettler, from the files of M. L. Schuster)

Fig. 105. Wildfire, caused by *Pseudomonas syringae* pv. *tabaci*, on the upper leaf surface. (Courtesy H. F. Schwartz)

gen cells block movement of water through the vascular system. Foliar symptoms often progress and appear as interveinal chlorosis and necrosis. Damage is particularly severe when young plants become infected. Seedlings are frequently stunted or killed. Water-soak inoculation of leaves produces white pustules (Fig. 102). Stem inoculation produces veinal yellowing and necrosis in leaves (Fig. 103). Infected seeds may show purple or yellow discoloration (Fig. 104). Seedlings from infected seeds develop purple discoloration of the stem.

Causal Organism

Bacterial wilt is caused by the bacterium *Curtobacterium flaccumfaciens* pv. *flaccumfaciens* (Hedges) Collins & Jones (syn. *Corynebacterium flaccumfaciens* subsp. *flaccumfaciens* (Hedges) Dowson). The bacterium is gram positive and is characterized by short pleomorphic rods. Snapping (bending) division is characteristic, and cells do not form spores. The bacterium is motile by one or a few polar or subpolar flagella. Aesculin is hydrolyzed. Several variants of the bacterium have been identified that produce orange and purple pigments.

The bacterium is infectious to other hosts in the family Leguminosae. Reported hosts are azuki bean, black gram, common bean, cowpea, mung bean, scarlet runner bean, and soybean. The wilt pathogen induces bacterial tan spot in soybean. *Zea mays* L. also has been reported to be susceptible after inoculation.

Disease Cycle and Epidemiology

The pathogen is carried on or in seeds and can overwinter in plant debris or on weeds. Survival is poor in soil. Initial infection occurs when the pathogen gains access to the vascular system either through systemic movement from infected seeds or through wounds on aboveground plant parts. Stomata are apparently infrequently penetrated, and disease probably does not develop from stomatal ingress. This contrasts with the common blight and halo blight bacteria, which routinely enter their hosts through stomata. Development of bacterial wilt is rapid following hailstorms and at temperatures higher than 32°C; the pathogen grows most rapidly at 37°C.

Management

Pathogen-free seeds and management measures recommended for other bacterial diseases of beans should be used for management of bacterial wilt.

Selected References

Collins, M. D., and Jones, D. 1983. Reclassification of *Corynebacterium flaccumfaciens, Corynebacterium betae, Corynebacterium oortii*, and *Corynebacterium poinsettiae* in the genus *Curtobacterium*, as *Curtobacterium flaccumfaciens* comb. nov. J. Gen. Microbiol. 129:3545-3548.

Harveson, R. M., Vidaver, A. K., and Schwartz, H. F. 2005. Bacterial wilt of dry beans in western Nebraska. NebGuide G05-1562-A. Cooperative Extension, Institute of Agriculture and Natural Resources, University of Nebraska, Lincoln.

Hsieh, T. F., Huang, H. C., Erickson, R. S., Yanke, L. J., and Mündel, H.-H. 2002. First report of bacterial wilt of common bean caused by *Curtobacterium flaccumfaciens* in western Canada. Plant Dis. 86:1275.

Hsieh, T. F., Huang, H. C., Mündel, H.-H., Conner, R. L., Erickson, R. S., and Balasubramanian, P. M. 2005. Resistance of common bean (*Phaseolus vulgaris*) to bacterial wilt caused by *Curtobacterium flaccumfaciens* pv. *flaccumfaciens*. J. Phytopathol. 153:245-249.

Rickard, S. F., and Walker, J. C. 1965. Mode of inoculation and host nutrition in relation to bacterial wilt of bean. Phytopathology 55:174-178.

Schuster, M. L., Vidaver, A. K., and Mandel, M. 1968. A purple pigment-producing bean wilt bacterium, *Corynebacterium flaccumfaciens* var. *violaceum*, n. var. Can. J. Microbiol. 14:423-427.

(Prepared by A. W. Saettler;
Revised by C. Ishimaru, S. K. Mohan, and G. D. Franc)

Wildfire

The bacterium *Pseudomonas syringae* pv. *tabaci* (Wolf & Foster) Young et al., a common pathogen of tobacco and soybean, causes wildfire disease of beans in Brazil. The bacterium has a wide host range and exhibits a high degree of pathogenic specialization among strains isolated from different hosts. The bean pathogen is a typical fluorescent pseudomonad, is oxidase negative, hydrolyzes aesculin, and produces a toxin (tabtoxin) in vitro, which induces wildfire symptoms in bean leaves. *P. syringae* pv. *tabaci* utilizes erythritol, sorbitol, and L-tartrate (but not DL-lactate) and is pectolytic.

Symptoms on beans occur only on leaves and consist of necrotic, brown, circular or angular lesions bordered by a pronounced yellow halo (Fig. 105). Lesions occasionally coalesce, resembling symptoms of common bacterial blight. Since pods apparently are not infected, seed infection is unlikely, but seed contamination (infestation) is likely.

Little is known regarding primary sources of inoculum or conditions under which the pathogen spreads; hence, no management measures are recommended.

Selected References

Mohan, S. K. 1984. Wildfire of dry beans caused by *Pseudomonas syringae* pv. *tabaci* (Wolf and Foster) Young et al. in the state of Parana, Brazil. Annu. Rep. Bean Improv. Coop. 27:108-109.

Ribeiro, R. de L. D., Hagedorn, D. J., Durbin, R. D., and Uchytil, T. F. 1979. Characterization of the bacterium inciting bean wildfire in Brazil. Phytopathology 69:208-212.

(Prepared by A. W. Saettler;
Revised by C. Ishimaru, S. K. Mohan, and G. D. Franc)

Diseases Caused by Nematodes

Many plant-parasitic nematodes are associated with roots and soils of beans, and several of these cause considerable damage and economic yield losses. However, species of *Meloidogyne* (root-knot) and *Pratylenchus* (lesion) nematodes are the most frequently encountered and reported to cause damage on beans. Large populations of these nematodes cause significant yield losses that may reach 80% with lesion nematodes and 90% with root-knot nematodes. In addition, plant-parasitic nematodes, particularly the root-knot nematodes, predispose many crop plants, including beans, to various physical and biotic stress factors, including soilborne fungal pathogens, that incite root rot and wilt diseases. They also alter a plant's metabolism such that a plant that would normally be resistant to a particular pathogen becomes susceptible to that pathogen.

Other plant-parasitic nematodes, including reniform (*Rotylenchulus reniformis* Linford & Oliveira), soybean cyst (*Heterodera glycines* Ichinohe), sting (*Belonolaimus longicaudatus* Rau), spiral (*Helicotylenchus dihystera* (Cobb) Sher), stunt (*Tylenchorhynchus* spp.), and ring (*Criconemoides ovantus* Raski), have been reported to feed on, reproduce on, and damage beans. However, these nematodes seldom occur at high population densities on beans, are not as prevalent as root-knot and lesion nematodes, or both. The root-knot and lesion nematodes are discussed in greater detail below.

Root-Knot Nematodes

Root-knot nematodes are worldwide in distribution and have an extensive host range, including agronomic crops and weeds representing many plant families. Yield losses resulting from damage by root-knot nematodes are most severe in light-textured soils with good drainage. Although quantitative data on bean yield losses caused by root-knot nematodes are limited, extensive losses have occurred in Latin America, southern regions of the United States, and the coastal region of Peru. Detailed information dealing with the distribution, ecology,

and management of root-knot nematodes (*Meloidogyne* spp.) can be found in publications of the International *Meloidogyne* Project (USAID/Department of Plant Pathology, North Carolina State University, Raleigh).

Symptoms

Aboveground symptoms exhibited by *Meloidogyne*-infected plants are nonspecific and do not permit positive diagnosis. Severely infected plants may appear chlorotic, stunted, necrotic, or wilted, especially during periods of moisture stress and high temperatures. Diagnostic symptoms appear on roots and hypocotyls of infected plants as galls or knots (Figs. 106–108). Galls are 1–10 mm in diameter or larger, depending on the nematode species involved and the location of the galls in the root system. Severely galled root systems become malformed with shortened and thickened individual roots so that roots may appear as a mass of galls. Infection with root-knot nematodes frequently suppresses root branching and the rate of root growth. The altered root growth reduces the volume and surface area of the roots and, consequently, restricts the uptake of water and minerals and the synthesis of cytokinins, gibberellins, and other growth-determining metabolites. Intensive galling seriously reduces root efficiency and may result in permanent wilting, premature defoliation, and eventually plant death.

Causal Organisms

Root-knot nematodes are sedentary, obligate endoparasites that have evolved specialized and complex relationships with their hosts. Of the more than 50 species of root-knot nematodes, four account for approximately 99% of populations collected from cultivated crop species, including beans. These are *M. incognita* (Kofoid & White) Chitwood, *M. javanica* (Treub) Chitwood, *M. hapla* Chitwood, and *M. arenaria* (Neal) Chitwood. Four races of *M. incognita* and two races of *M. arenaria* have been identified. Host races have not been identified among natural populations of *M. hapla* and *M. javanica*. *M. hapla* occurs in relatively cold areas and survives at temperatures as low as –15°C. The other three major species of root-knot nematodes are adapted to,

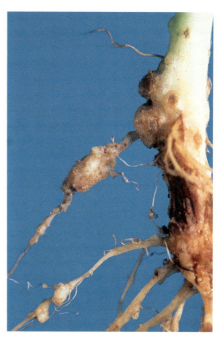

Fig. 106. Galls on cultivar Provider bean roots caused by the root-knot nematode, a *Meloidogyne* sp. (Courtesy G. S. Abawi, from the files of G. Fassuliotis)

Fig. 107. Galls on bean roots and hypocotyl caused by the root-knot nematode, a *Meloidogyne* sp. (Courtesy H. F. Schwartz, from the files of M. A. Pastor-Corrales)

Fig. 108. Galls on hypocotyls and adventitious roots caused by the root-knot nematode, a *Meloidogyne* sp. (Courtesy G. S. Abawi)

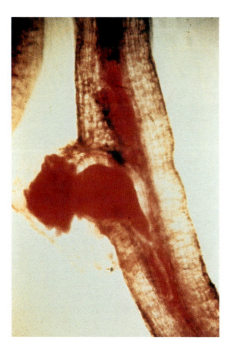

Fig. 109. Mature *Meloidogyne* female attached to root with a full egg sac. (Courtesy W. F. Mai)

and occur in, high-temperature areas. The most prevalent species of root-knot nematodes in tropical and subtropical regions are *M. incognita* and *M. javanica*.

Root-knot nematodes survive in soil as eggs and juveniles. Eggs are deposited by mature females in an egg sac consisting of a gelatinous matrix that protects eggs from dehydration (Fig. 109). As many as 1,000 eggs or more may be deposited per egg sac. The length of survival of root-knot nematodes in soil varies with the species, stage of development, soil texture, soil moisture, soil aeration, and other factors. Nematodes may be

disseminated in irrigation water, vegetative plant parts, and infested soil adhering to farm implements, animals, and people.

Disease Cycle

The life cycle of *Meloidogyne* spp. and other plant-parasitic nematodes involves five developmental stages (egg plus four juvenile stages) before differentiation into male or female adults (Fig. 110). Embryogenic development within eggs results in the formation of vermiform first-stage juveniles that later molt into second-stage juveniles (J2). Eggs are oval to ellipsoidal, slightly concave, and 30–52 × 67–128 μm. The J2s are 15 μm wide and 375–500 μm long; they hatch by breaking the egg shell with repeated thrusting of their well-developed stylet (about 10 μm long). Hatching may occur without root stimuli, although root diffusates have been reported to increase egg hatch. J2s, the only infective stage of the nematode, migrate through soil and are attracted to roots. They penetrate the region directly behind the root cap by thrusting their stylets and possibly through enzymatic activities. Once inside, J2s migrate intercellularly through cortical tissues toward the region of cell differentiation. There they become sedentary with their heads inserted in the periphery of vascular tissues (primary phloem or xylem parenchyma cells) and the rest of their bodies positioned in the cortex parallel to the long axis of the root.

In susceptible hosts, a highly specialized cellular adaptation occurs around the nematode head, resulting in the formation of permanent feeding sites called giant cells. Feeding and secretions by the nematode induce plant cells in the vicinity to increase in number (hyperplasia) and size (hypertrophy), resulting in swellings called galls or knots. The sedentary juveniles continue to enlarge during the formation of giant cells and galls, completing the second and third molts, after which the sexes can be differentiated. The fourth molt produces mature males and females. Adult males are cylindroid (1.2–1.5 mm long), lack a bursa, and have a well-developed stylet. They have no reproductive function in most *Meloidogyne* species, but *M. hapla* is one of the exceptions. Mature females are pyriform (0.27–0.75 × 0.4–1.3 mm) and pearly white; they have soft cuticles and can be seen on the surface of galled tissues after rupturing the root cuticle (Fig. 109). The life cycle may be completed in 17–57 days, depending on many factors, especially soil temperature.

Epidemiology

Moisture status, aeration, texture, and temperature of the soil are major physical factors that affect the distribution and reproduction of root-knot nematodes and the damage they cause to host crops. J2s migrate through soil water films and are most active at soil moisture levels of 40–60% of field capacity. Nematode activity generally decreases in drier and wetter soils. Eggs are killed by exposure to anaerobic conditions of saturated soils. Temperature plays a major role in hatching, host infection, reproduction, and survival of root-knot nematodes and thus affects their geographic distribution. Species of root-knot nematodes react differently to temperature, particularly low temperatures. Tolerance of low temperatures among the four major species decreases in the following order: *M. hapla*, *M. incognita*, *M. arenaria*, and *M. javanica*. Root-knot nematodes are most abundant, and cause the most serious damage to host crops (including beans), in sandy soils with good drainage. Very few populations of root-knot nematodes have been found in soils with more than 40% clay or 50% silt fractions. In addition, migration and penetration of host plants by root-knot nematodes decrease as the clay and silt fractions increase from 14 to 33%. Plants stressed by physical or biological factors are more sensitive to damage by these nematodes.

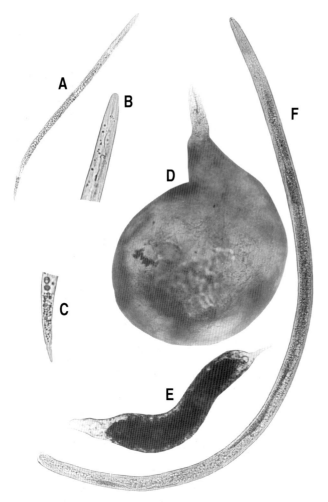

Fig. 110. Life stages of a *Meloidogyne* sp.: **A,** second-stage juvenile (×119); **B,** anterior portion of second-stage juvenile (×300); **C,** tail of second-stage juvenile (×330); **D,** mature female (×133); **E,** developing juvenile from plant root (×133); and **F,** mature male (×119). (Courtesy W. F. Mai and H. H. Lyon)

Management

Application of the principles and practices of integrated pest management is essential to maintain root-knot nematodes below damaging threshold levels and to reduce the chances of their introduction to noninfested areas. Considerable information is available on the costs and benefits of different options for management of root-knot nematodes on beans.

While crop rotation is an effective management measure for many soilborne nematodes, the extensive host range of root-knot nematodes makes it difficult to select an appropriate rotation. Nevertheless, densities of root-knot nematodes can be reduced considerably by planting beans once every two or three seasons in rotation with nonhosts, such as grasses. Growing crops that are antagonistic to nematodes, such as marigold (*Tagetes minuta* L.), rattlebox (*Crotalaria spectabilis* Roth), sorghum-sudangrass hybrids (*Sorghum bicolor* (L.) Moench × *S. sudanense* (Piper) Stapf.), hairy indigo (*Indigofera hirsuta* L.), bahiagrass (*Paspalum notatum* Flueggé), and velvet bean (*Mucuna deeringiana* (Bort.) Merr.), can reduce populations of plant-parasitic nematodes. Long fallow periods, deep plowing, weed management, and flooding are also effective in reducing nematode populations and can be used where practical. Several biological control agents have been experimentally shown to be effective parasites of eggs and adult stages of root-knot nematodes. In field tests, the fungus parasite *Paecilomyces lilacinus* (Thom) R. A. Samson has attacked eggs of root-knot and other nematodes, reduced nematode densities, and protected beans from damage.

Use of highly resistant or tolerant bean cultivars, where available, is the most effective and economical method for managing root-knot nematodes. Numerous reports are available in the literature that deal with the evaluation of bean cultivars and the identification of breeding lines with resistance to these nematodes. However, resistance to one species or one race of root-knot nematodes may be independent of resistance to other species or races. For example, P.I. 165426 was reported to be resistant to *M. incognita* but susceptible to simultaneous inoculation with *M. incognita* and *M. javanica*. The resistance may be governed by multiple gene action. Sources of resistant bean germ plasm include Nemasnap, Tendergreen, Tenderpod, Saginaw, Wingard Wonder, Manoa Wonder, Carioca, Ex Rico 23, P.I. 165435, P.I. 313709, A107, A211, A315, A445, G1805, G2618, Alabama 1, Alabama 2, Alabama 8, and Alabama 19.

Management of root-knot nematodes with nematicides can be effective, and nematicides are widely used in the production of many agronomic crops. However, they are very expensive and may not be justified for a crop such as beans. Nematicides must be handled carefully and often require the use of special application equipment. Fumigant-type nematicides, such as methyl bromide, chloropicrin, methyl isothiocyanate, D-D (1,3-dichloropropene and related hydrocarbons), and others, have been used successfully to manage these nematodes on beans and other crops. Management of root-knot nematodes and increase in bean yield have also been obtained with the use of non-fumigant-type nematicides, including fenamiphos, carbofuran, oxamyl, and aldicarb.

Selected References

France, R. A., and Abawi, G. S. 1994. Interaction between *Meloidogyne incognita* and *Fusarium oxysporum* f. sp. *phaseoli* on selected bean genotypes. J. Nematol. 26:467-474.

Melakeberham, H., Webster, J. M., Brooke, R. C., and D'Auria, J. M. 1987. Effect of *Meloidogyne incognita* on plant nutrient concentration and its influence on the physiology of beans. J. Nematol. 19:324-330.

Mullin, B. A., Abawi, G. S., Pastor-Corrales, M. A., and Kornegay, J. L. 1991. Root-knot nematodes associated with beans in Colombia and Peru and related yield loss. Plant Dis. 75:1208-1211.

Mullin, B. A., Abawi, G. S., Pastor-Corrales, M. A., and Kornegay,

J. L. 1991. Reactions of selected bean pure lines and accessions to *Meloidogyne* species. Plant Dis. 75:1212-1216.

Omwega, C. O., Thomason, I. J., and Roberts, P. A. 1990. A single dominant gene in common bean conferring resistance to three root-knot nematode species. Phytopathology 80:745-748.

Sydenham, G. M., McSorley, R., and Dunn, R. A. 1996. Effect of resistance in *Phaseolus vulgaris* on development of *Meloidogyne* species. J. Nematol. 28:485-491.

Widmer, T. L., and Abawi, G. S. 2000. Mechanism of suppression of *Meloidogyne hapla* and its damage by a green manure of sudan grass. Plant Dis. 84:562-568.

(Prepared by G. S. Abawi, B. A. Mullin, and W. F. Mai;
Revised by G. S. Abawi)

Lesion Nematodes

Lesion nematodes (*Pratylenchus* spp.), also referred to as meadow nematodes, are migratory endoparasites and, unlike root-knot nematodes, are vermiform during all developmental stages. They are distributed worldwide but occur at high density most frequently in temperate regions. In the northeastern region of the United States, lesion nematodes are considered to be the most important plant-parasitic nematodes on crop plants. Damage to beans by lesion nematodes depends on many factors, including initial population densities of the nematodes. Under greenhouse conditions, growth of susceptible bean cultivars was reduced by 50 or more *P. penetrans* (Cobb) Filip. & Schuur.-Stek. per 100 cm³ of soil. Yield of susceptible cultivars was reduced by 43–47% at densities of 150 *P. penetrans* per 100 cm³ of soil. Damage to beans by lesion nematodes may also result from interaction with other soilborne microorganisms infecting roots. For example, incidence and severity of Fusarium root rot on beans may increase in the presence of *P. penetrans*. Many plant species are hosts of lesion nematodes but differ in the degree to which they support nematode development and reproduction. Lesion nematodes survive as eggs, larvae, or adults free in soil or in tissues of host crops in soil. They are spread by infested soil adhering to farm implements, machinery, animals, and people, and by wind, water, and infected or contaminated propagating plant materials.

Symptoms

Bean plants heavily infected with lesion nematodes have poorly developed roots that may bear small, brown-black lesions (Fig. 111). Lesions result from nematode penetration and feeding activities in epidermal and cortical tissues (Fig. 112)

and may coalesce and kill small fibrous roots. Diagnostic proof of lesion nematode infections requires extraction of these nematodes from root tissues and surrounding soil. Above-ground symptoms of severely infected plants include low vigor, stunting, chlorosis, and wilting. Destruction of roots by lesion nematodes is more severe in the presence of soilborne pathogens and certain saprophytic organisms.

Causal Organisms

There are about 40 reported *Pratylenchus* species, but three are most frequently found in association with beans. These are *P. penetrans*, *P. brachyurus* (Godfrey) Filip. & Schuur.-Stek., and *P. scribneri* Steiner. Other species reported infecting or associated with beans include *P. neglectus* (Rensch) Filip. & Schuur.-Stek., *P. thornei* Sher & Allen, *P. zeae* Graham, *P. goodeyi* Sher & Allen, and *Paratylenchus projectus* Jenkins. All stages of these nematodes are vermiform (cylindroid) and less than 1 mm long. Microscopic diagnostic features (Fig. 113) include the overlapping esophagus, blunt tail, flat head, and slow, graceful movement. Males and females are alike except for the sexual organs. Males are common and required for reproduction in some species. Lesion nematodes are more numerous in temperate than in tropical or subtropical areas, and some species appear to be adapted better than others to cool regions.

Disease Cycle

All life stages of lesion nematodes can penetrate roots, but adult females and fourth-stage juveniles enter roots more frequently and thus are the most important infective stages. Gravid females lay eggs singly or in clusters within cortical tissues of roots. The first molt occurs in the egg, resulting in the second-stage juvenile (J2). The J2 hatches from the egg and molts three more times to produce the adult female or male. Juveniles migrate throughout cortical tissues of roots and below-ground parts, feeding on and damaging or killing cells. Lesion nematodes move to new roots upon overcrowding or decay of infected roots. The length of the life cycle ranges from 25 to 50 days, depending on the nematode species, host species, and many other physical and biological factors.

Fig. 111. Discoloration and reduction of root and hypocotyl tissues (right) caused by the lesion nematode *Pratylenchus penetrans*. Healthy root and hypocotyl tissues are on the left. (Courtesy G. S. Abawi)

Fig. 112. *Pratylenchus penetrans* lesions on the root of a lima bean plant. (Courtesy G. S. Abawi)

Epidemiology

Lesion nematodes survive best in cool (10–15°C), moist soils. The life cycle is shortest at 30°C, and high temperatures usually increase the ratio of adult males to females. Lesion nematodes can survive for only a few hours at −8 to −12°C when they are free in soil as juveniles or adults. They are most common and most damaging in sandy soils, where their reproduction is greatest. Damage by lesion nematodes is especially severe when soil moisture or nutrients are limited. Also, reproduction of lesion nematodes is highest at low to neutral soil reaction (pH 5.2–6.4). Addition of a large amount of organic matter tends to reduce the populations of lesion nematodes, possibly as a result of increased activities of parasites and predators.

Management

Effective management of lesion nematodes can be achieved by integrated measures that prevent populations from increasing to damaging levels. Lesion nematodes have an extensive host range, thus their populations are generally affected little by crop rotation. However, agronomic plants differ in their efficiency as hosts and in the effect of their incorporated green manures against the lesion nematodes. It has been reported that the soil population of lesion nematodes may be reduced by growing certain cover crops (such as oats [cultivar Saia], marigold [cultivars Polynema and Nema-gone], *Sesbania sesban* (L.) Merr., *Tephrosia vogelii* Hook f., and certain cultivars of ryegrass, sorghum-sudangrass hybrids, and rapeseed, as well as other plants), by incorporating their green manures, or both. Summer fallow or flooding also reduces populations, especially in areas with hot, dry climates.

Limited information is available on the reaction of bean cultivars to lesion nematodes. The navy bean cultivars Saginaw, Seafarer, and Tuscola were reported to be tolerant of *P. penetrans*. Great northern 1140 and Idaho 11 were reported resistant to both *P. neglectus* and *P. hexincisus* Taylor & Jenkins.

Chemical management of lesion nematodes and increase of bean yields have been obtained with fumigant (e.g., D-D [1,3-dichloropropene and related hydrocarbons], methyl bromide, methyl isothiocyanate, and chloropicrin) and nonfumigant (e.g., fenamiphos, oxamyl, carbofuran, and aldicarb) nematicides broadcast over the field or applied along the row. Oxamyl applied to foliage has also provided good management. However, these nematicides are not easy to apply and may not be economical to use.

Selected References

Abawi, G. S., and Widmer, T. L. 2000. Impact of soil health management practices on soil-borne pathogens, nematodes and root diseases of vegetable crops. Appl. Soil Ecol. 15:37-47.

Davis, E. L., and MacGuidwin, A. E. 2000. Lesion nematode disease. The Plant Health Instructor. American Phytopathological Society, St. Paul, MN. DOI: 10.1094/PHI-I-2000-1030-02.

Elliott, A. P., and Bird, G. W. 1985. Pathogenicity of *Pratylenchus penetrans* to navy bean (*Phaseolus vulgaris* L.). J. Nematol. 17:81-85.

Mai, W. F., Bloom, J. R., and Chen, T. A. 1977. Biology and ecology of the plant-parasitic nematode *Pratylenchus penetrans*. Penn. State Univ. Agric. Exp. Stn. Bull. 815.

Rich, J. R., Keen, K. T., and Thomason, I. J. 1977. Association of coumestans with the hypersensitivity of lima bean roots to *Pratylenchus scribneri*. Physiol. Plant Pathol. 10:105-116.

Robbins, R. T., Dickerson, O. J., and Kyle, J. H. 1972. Pinto bean yield increased by chemical control of *Pratylenchus* spp. J. Nematol. 4:28-32.

Thomason, I. J., Rich, J. R., and O'Melia, F. C. 1976. Pathology and histopathology of *Pratylenchus scribneri* infecting snap bean and lima bean. J. Nematol. 8:347-352.

(Prepared by G. S. Abawi, B. A. Mullin, and W. F. Mai; Revised by G. S. Abawi)

Reniform Nematode

Damage to snap beans caused by the reniform nematode (*Rotylenchulus reniformis* Linford & Oliveira) has been reported in sandy soils naturally infested with this nematode in southern Florida. Bean yield was highly correlated with the final but not the initial population density of *R. reniformis*. Soil incorporation of velvet bean manure reduced the population of *R. reniformis* (also *Meloidogyne incognita* (Kofoid & White) Chitwood) and greatly increased the yield of the following dry bean crop. Nematicide treatments increased bean yield when the population of reniform nematodes was more than 400/cm^3 of soil, but effects on yield were inconsistent at lower population densities.

Selected References

Abawi, G. S., and Varon de Agudelo, F. 1989. Nematodes. Pages 433-453 in: Bean Production Problems in the Tropics, 2nd ed. H. F. Schwartz and M. A. Pastor-Corrales, eds. Centro Internacional de Agricultura Tropical (CIAT), Cali, Colombia.

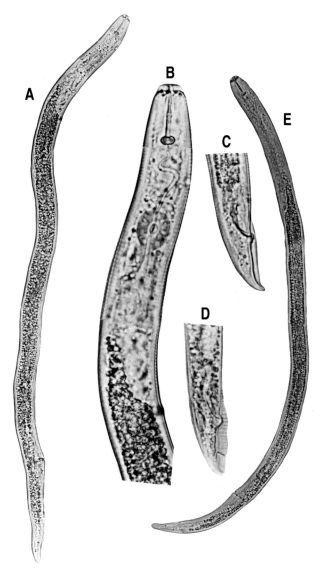

Fig. 113. Life stages of a *Pratylenchus* sp.: **A,** mature female (×230); **B,** anterior portion of female illustrating the overlapping esophagus (×610); **C and D,** lateral view of male tail (×413); **E,** mature female (×230); and **F,** mature male (×119). (Courtesy W. F. Mai and H. H. Lyon)

McSorley, R. 1980. Effect of *Rotylenchulus reniformis* on snap bean and methods for control by oxamyl. Nematropica 10:89-95.

(Prepared by G. S. Abawi, B. A. Mullin, and W. F. Mai; Revised by G. S. Abawi)

Soybean Cyst Nematode

Many dry and snap bean cultivars support growth and development of the soybean cyst nematode, *Heterodera glycines* Ichinohe. Invasion, juvenile development, length of life cycle, and final population densities of soybean cyst nematodes on beans may equal or exceed those on susceptible soybean cultivars. Infections of commercially grown beans with *H. glycines* have been reported. However, information on the effect of the soybean cyst nematode on yield of snap and dry beans is very limited. Additionally, it was reported that there was no effect of soybean cyst nematode on the severity of Fusarium root rot of beans. The bean lines WIS 36, WIS 46, WIS 147, and the Brazilian line L-2300 were reported as resistant to *H. glycines*.

Selected References

Abawi, G. S., and Jacobsen, B. J. 1984. Effect of initial inoculum densities of *Heterodera glycines* on growth of soybean and kidney bean and their efficiency as hosts under greenhouse conditions. Phytopathology 74:1470-1474.

Abawi, G. S., and Varon de Agudelo, F. 1989. Nematodes. Pages 433-453 in: Bean Production Problems in the Tropics, 2nd ed. H. F. Schwartz and M. A. Pastor-Corrales, eds. Centro Internacional de Agricultura Tropical (CIAT), Cali, Colombia.

Melton, T. A., Noel, G. R., Jacobsen, B. J., and Hagedorn, D. J. 1985. Comparative host suitabilities of snap beans to the soybean cyst nematode (*Heterodera glycines*). Plant Dis. 69:119-122.

(Prepared by G. S. Abawi, B. A. Mullin, and W. F. Mai; Revised by G. S. Abawi)

Sting Nematode

Severe damage to beans has been reported in sandy soils infested with the sting nematode (*Belonolaimus longicaudatus* Rau). Nematicide treatment or a summer cover crop of hairy indigo (*Indigofera hirsuta* L.) greatly reduced the sting nematode population density and increased bean yield.

Selected References

Abawi, G. S., and Varon de Agudelo, F. 1989. Nematodes. Pages 433-453 in: Bean Production Problems in the Tropics, 2nd ed. H. F. Schwartz and M. A. Pastor-Corrales, eds. Centro Internacional de Agricultura Tropical (CIAT), Cali, Colombia.

Rhoades, H. L. 1974. Comparison of two methods of applying granular nematicides for control of sting nematode on snap beans, sweet corn and field corn. Soil Crop Sci. Soc. Fla. Proc. 33:77-80.

Rhoades, H. L. 1976. Effect of *Indigofera hirsuta* on *Belonolaimus longicaudatus*, *Meloidogyne incognita*, and *M. javanica* and subsequent crop yields. Plant Dis. Rep. 60:384-386.

(Prepared by G. S. Abawi, B. A. Mullin, and W. F. Mai; Revised by G. S. Abawi)

Other Nematodes

Other plant-parasitic nematodes have been found in association with bean soils and roots, but their potential to damage beans is generally unknown. These nematodes include *Paratrichodorus christie* (Allen) Siddiqi, *Dolichodorus heterocephalus* Cobb, *Criconemoides ovantus* Raski, *Helicotylenchus dihystera* (Cobb) Sher, *Quinisulcius acutus* (Allen) Siddiqi, *Hoplolaimus* spp., *Longidorus* spp., *Tylenchorhynchus* spp., *Xiphinema* spp., and *Ditylenchus* spp.

The chrysanthemum foliar nematode, *Aphelenchoides ritzemabosi* (Schwartz) Steiner & Buhrer, is the causal agent of nematode angular leaf spot, a foliar disease of pinto bean described by Franc et al. Nematode angular leaf spot is typified by numerous dark, angular lesions on leaves and, occasionally, a superficial necrosis on the upper surface of the petiole (Fig. 114). Expansion of individual lesions was limited by leaf veins, with most lesions ranging in size from several millimeters to 1 cm. *Aphelenchoides ritzemabosi* readily parasitizes alfalfa and routinely is found in association with the alfalfa stem nematode, *Ditylenchus dipsaci* (Kühn) Filip. in Wyoming and other western states. Crop rotation with nonhost crops reduces the presence of inoculum. Pinto bean rotation with alfalfa may increase inoculum and disease development.

Selected References

Abawi, G. S., and Varon de Agudelo, F. 1989. Nematodes. Pages 433-453 in: Bean Production Problems in the Tropics, 2nd ed. H. F. Schwartz and M. A. Pastor-Corrales, eds. Centro Internacional de Agricultura Tropical (CIAT), Cali, Colombia.

Franc, G. D., Beaupré, C. M.-S., Gray, F. A., and Hall, R. D. 1996. Nematode angular leaf spot of dry bean in Wyoming. Plant Dis. 80:476-477.

Johnson, A. W., Jaworski, C. A., Sumner, D. R., and Chalfant, R. B. 1979. Effects of film mulch, trickle irrigation, and soil pesticides on nematodes and yield of polebean. Plant Dis. Rep. 63:360-364.

(Prepared by G. S. Abawi, B. A. Mullin, and W. F. Mai; Revised by G. S. Abawi)

Fig. 114. Leaf lesions of nematode angular leaf spot, caused by *Aphelenchoides ritzemabosi*. (Courtesy G. D. Franc)

Diseases Caused by Viruses

Common bean is very susceptible to plant viruses, under both natural and experimental conditions. Hence, viral diseases are a major constraint to common bean production wherever this legume is grown. Most of the viruses that infect common bean under natural conditions do not seem to have evolved with this plant species in its center of origin, the Americas. Those viruses that were first detected in the center of origin have a closer biological relationship with common bean landraces; are generally seed transmitted; and do not cause major damage to their host. On the contrary, when common bean encounters exotic viruses, significant yield losses often are a result. In recent decades, however, different opportunistic viruses, capable of species jumping, have also become major pathogens of common bean. These viruses have arisen as a result of agricultural expansionism, implementation of mixed cropping systems, and emergence of large populations of insect vectors.

Proper identification of viral pathogens is necessary to implement effective integrated disease management practices. The viral diseases described in this section should contribute to the correct identification of the main viruses that infect common bean worldwide. Some diseases are given a generic name when various related but different viruses have been shown to induce similar symptoms. The illustrations selected here represent the most common symptoms associated with the viral diseases described; however, symptom expression is conditioned by many genetic and environmental factors. Complementary information is also provided in the selected references cited after each individual virus description. Further characterization of common bean and other viruses infecting this legume can be accomplished by using traditional and modern diagnostic techniques available at many agricultural research institutions that work with common bean, as well as from an increasing number of commercial companies offering virus identification materials and diagnostic kits.

Alfalfa Mosaic

Alfalfa mosaic was first described in 1931 as a viral disease of alfalfa (*Medicago sativa* L.) in California. In 1949, a yellow dot disease of common bean was shown to be caused by the causal agent of alfalfa mosaic in Washington. Alfalfa mosaic is distributed worldwide, primarily in temperate regions where alfalfa, clover, and other temperate legume pastures are grown. Different strains of *Alfalfa mosaic virus* (AMV) have been shown to naturally infect common bean in the western United States, Canada, eastern Germany, northern Italy, and Chile in South America.

Symptoms

Characteristic symptoms induced by AMV in common bean include intense foliar yellowing, yellow mottling, leaf and pod malformation, and stem and pod necrosis (Fig. 115).

Causal Agent

AMV is the only species known of the genus *Alfamovirus* of the family *Bromoviridae*. Virions are 18 nm wide and 18–56 nm long and range in shape from spherical to bacilliform. The particles encapsidate three major species of linear, single-stranded, positive-sense RNA with molecular sizes of 3.6, 2.6, and 2.0 kb. The coat protein subunit is expressed from a subgenomic RNA and has a molecular weight of approximately 24,000. For infectivity, RNAs 1, 2, and 3 are required, plus either RNA 4 or the coat protein. Some AMV genes are homologous with those of the genus *Ilarvirus* of the family *Bromoviridae*. AMV has been reported to infect more than 232 plant species in 48 families, including 54 species of legumes.

Disease Cycle and Epidemiology

The primary means of AMV dissemination around the world is by seeds produced from systemically infected alfalfa plants (up to 50% seed transmission). Plant-to-plant spread of AMV under field conditions is by aphids. The virus is transmitted in a nonpersistent manner by more than 14 aphid species, including *Acyrthosiphon pisum* Harris (pea aphid), *Aphis fabae* Scopoli (bean aphid), and *Myzus persicae* Sulzer (green peach aphid). AMV is also transmitted by pollen to the developing seeds and by mechanical means. Seed transmission of AMV has been observed in beans inoculated with a strain of the virus from *Amaranthus albus* L. from eastern Washington. The virus was transmitted in 0.7–4.9% of the seeds in 3 of 12 bean lines and in 15.5% of the seeds of *A. albus*. This strain, when seedborne, did not produce symptoms in infected bean plants. Transmission of AMV was not observed in seeds of bean cultivars infected systemically with different strains of the virus from the Pacific Northwest of the United States.

Several annual and perennial food, forage, ornamental, and weed species have been identified as natural hosts of different strains of AMV. Many of these hosts serve as important reservoirs of the virus. In many regions of the world, alfalfa is the primary reservoir of AMV.

Management

Establishment of bean plantings close to alfalfa, clover, or other important legume reservoirs of AMV should be avoided. Seed transmission of AMV in common bean is usually very low. Fortunately, AMV does not seem to be well adapted to common bean despite its capacity to infect most *Phaseolus vulgaris* L. genotypes inoculated to date.

Selected References

Jaspars, E. M. J., and Bos, L. 1980. Alfalfa mosaic virus. Descriptions of Plant Viruses, No. 229. Commonwealth Mycological Institute and Association of Applied Biologists, Kew, Surrey, England.

Kaiser, W. J., and Hannan, R. M. 1983. Additional hosts of alfalfa mosaic virus and its seed transmission in tumble pigweed and bean. Plant Dis. 67:1354-1357.

Sepulveda, P., Morales, F., and Castaño, M. 2001. Detection of alfalfa mosaic virus in bean (*Phaseolus vulgaris* L.) production regions of Chile. Agric. Téc. 61:379-384.

Thomas, H. R. 1951. Yellow dot, a virus disease of bean. Phytopathology 41:967-974.

Fig. 115. Systemic yellow spotting of leaves caused by *Alfalfa mosaic virus*. (Courtesy F. J. Morales)

Thomas, H. R. 1953. Isolation of alfalfa mosaic virus strains from field-grown beans. Plant Dis. Rep. 37:390-391.

Zaumeyer, W. J. 1963. Two new strains of alfalfa mosaic virus systemically infectious to bean. Phytopathology 53:444-449.

Zaumeyer, W. J., and Patino, G. 1960. Vein necrosis, another systemically infectious strain of alfalfa mosaic virus in bean. Phytopathology 50:226-231.

(Prepared by W. J. Kaiser and R. M. Hannan; Revised by F. J. Morales)

Angular Mosaic

Angular mosaic is a restricted disease of common bean in southeastern Brazil, mainly in the states of São Paulo and Paraná, but its incidence there may be high in common bean fields, as well as in soybean and cowpea. In Africa, the disease has also been reported to affect common bean in Tanzania and, possibly, in Egypt. The causal virus spreads in other legume hosts, such as mung bean, in Asia, Africa, and the Pacific.

Symptoms

The most characteristic symptoms are yellow, angular leaf spots against the normal green background. These symptoms, however, are only observed in certain common bean genotypes, such as the Jalo and Manteiga grain types. In most other bean cultivars grown in Brazil, infected plants remain symptomless. Symptoms usually start as veinclearing and mild mosaic, increasing in intensity as the infected bean leaves mature. The number of pods per plant and seeds per pod are generally reduced in common bean plants affected by angular mosaic.

Causal Agent

The causal agent of bean angular mosaic is *Cowpea mild mottle virus* (CPMMV), a species of the genus *Carlavirus*. Virions are flexuous filaments, approximately 650 nm in length and 13 nm in diameter, and contain a single-stranded RNA molecule approximately 7.5 kb in size. The capsid protein subunit has a relative molecular weight of 35,700.

Disease Cycle and Epidemiology

CPMMV is seedborne in some legumes other than common bean, and it can be mechanically transmitted. The natural vector of the virus is the whitefly species *Bemisia tabaci* (Gennadius). The biological transmission of CPMMV takes only a few minutes; hence, chemical management with contact pesticides is not effective. The whitefly vector usually reproduces in selected hosts, such as soybean in Brazil. Thus, this disease is frequently associated with large populations of *B. tabaci* in the affected localities.

Management

No specific management measures are known, other than those recommended for other whitefly-transmitted viruses, such as *Bean golden mosaic virus*. The basic strategy is not to plant susceptible common bean cultivars close to good reproductive hosts of the whitefly vector. Should this disease become important in the future, there are sources of genetic resistance in *Phaseolus vulgaris* L.

Selected References

Costa, A. S. 1987. Fitoviroses do feijoeiro no Brasil. Pages 173-256 in: Feijão: Fatores de Produção e Qualidade. E. A. Bulisani, ed. Fundacão Cargill, Campinas, São Paulo, Brazil.

Badge, J., Brunt, A., Carson, R., Dagless, E., Karamagioli, M., Phillips, S., and Seal, S. 1996. A carlavirus-specific PCR primer and partial nucleotide sequence provides further evidence for the recognition of cowpea mild mottle virus as a whitefly-transmitted carlavirus. Eur. J. Plant Pathol. 102:305-310.

Mink, G. I., and Keswani, C. L. 1987. First report of cowpea mild mottle virus on bean and mung bean in Tanzania. Plant Dis. 71:557.

(Prepared by F. J. Morales)

Bean Calico Mosaic

Bean calico mosaic was observed to affect common bean in northwestern Mexico in the early 1970s. At that time, the disease was thought to be caused by *Bean yellow mosaic virus*, but it was later associated with outbreaks of the whitefly *Bemisia tabaci* (Gennadius). Hence, it was assumed to be bean golden mosaic. It was not until the early 1990s that the causal agent was shown to be a different whitefly-transmitted geminivirus, and the name of this disease was changed to bean calico mosaic.

Symptoms

Initially, bean calico mosaic symptoms resemble those induced by bean golden mosaic and bean golden yellow mosaic, but infected plants tend to exhibit a noticeable chlorosis, bleaching, and subsequent necrosis of the foliage as the infection progresses (Fig. 116). Highly susceptible common bean genotypes also tend to abort their flowers or produce pods that are empty, distorted, or both.

Causal Agent

Bean calico mosaic virus (BCaMV) is recognized as a distinct species of the genus *Begomovirus* of the family *Geminiviridae*. However, BCaMV is most likely a pathogenic variant of the begomovirus *Squash leaf curl virus* (SLCV), which adapted to common bean in northwestern Mexico. In fact, SLCV is occasionally isolated in this region from common bean plants showing the characteristic chlorosis and yellowing associated with bean calico mosaic. As with most begomoviruses, BCaMV has a bipartite genome (DNA-A and DNA-B) each approximately 2.6 kb in size. The geminate virions consist of two incomplete icosahedra, approximately 18 × 30 nm. BCaMV is also related to *Cabbage leaf curl virus* (CaLCuV).

Disease Cycle and Epidemiology

BCaMV is transmitted in a persistent manner by different biotypes of the whitefly *B. tabaci*, particularly during the dry and warm months of the year. The intensive and continuous mixed cropping systems found in the desert southwest of the United States and northwestern Mexico favor the development of large populations of *B. tabaci* and transmission of different

Fig. 116. Foliar symptoms of bean calico mosaic. (Courtesy F. J. Morales)

begomoviruses found in horticultural crops throughout that region. The intensive use of pesticides has also led to the emergence of insecticide-resistant populations of the whitefly vector, which complicates management strategies.

Management

The management of BCaMV has been primarily achieved through the use of tolerant common bean cultivars that yield well, despite their systemic infection by BCaMV. The best sources of resistance to BCaMV in *Phaseolus vulgaris* L. have come from the Andean center of origin of this legume species. The isolation of common bean plantings from cucurbits harboring SLCV strains may be advisable to prevent opportunistic infections of common bean plants by this begomovirus.

Selected References

Brown, J. K., Chapman, M. A., and Nelson, M. R. 1990. Bean calico mosaic, a new disease of common bean caused by a whitefly-transmitted geminivirus. Plant Dis. 74:81.

Brown, J. K., Ostrow, K. M., Idris, A. M., and Stenger, D. C. 1999. Biotic, molecular, and phylogenetic characterization of bean calico mosaic virus, a distinct *Begomovirus* species with affiliation in the squash leaf curl virus cluster. Phytopathology 89:273-280.

Loniello, A. O., Martinez, R. T., Rojas, M. R., Gilbertson, R. L., Brown, J. K., and Maxwell, D. P. 1992. Molecular characterization of bean calico mosaic geminivirus. (Abstr.) Phytopathology 82:1149.

Morales, F. J., and Anderson, P. K. 2000. The emergence and dissemination of whitefly-transmitted geminiviruses in Latin America. Arch. Virol. 146:415-441.

(Prepared by F. J. Morales)

Bean Common Mosaic

Bean common mosaic was first recognized as a viral disease of common bean in the United States in 1917. The causal agent was first called bean mosaic virus, but the epithet "common" was added later to distinguish this virus from *Bean yellow mosaic virus*. The names common bean mosaic virus, bean mosaic virus, bean virus 1, and Phaseolus virus 1 have been used as synonyms.

Bean common mosaic reduces yield by as much as 80% and also adversely affects seed quality. Plants found to be naturally infected are common bean, *Phaseolus vulgaris* L. var. *aborigineus* (Burkart) Baudet, *Rhynchosia minima* (L.) DC., and some wild tropical species of the genus *Phaseolus*. Most common bean landraces in the American centers of origin of this legume species are susceptible to *Bean common mosaic virus* (BCMV).

Symptoms

Light and dark green mosaic with or without malformation (Figs. 117 and 118), leaf roll (Fig. 119), or yellow dots may be produced, often causing growth reduction. Some genotype–strain interactions produce a very mild mosaic or no symptoms at all in infected leaves.

Causal Agents

Bean common mosaic is caused by two different species of the genus *Potyvirus*, *Bean common mosaic virus* (BCMV) and *Bean common mosaic necrosis virus* (BCMNV). These viruses have flexuous filaments, approximately 15×750 nm, containing a single molecule of single-stranded RNA approximately

Fig. 118. Mosaic (left) and leaf distortion (right) caused by *Bean common mosaic virus*. (Courtesy J. R. Stavely, from the files of W. J. Zaumeyer)

Fig. 117. Typical green-on-green veinbanding symptoms of bean common mosaic on a trifoliolate leaf. (Courtesy E. Drijfhout)

Fig. 119. Leaf roll caused by *Bean common mosaic virus*. (Courtesy E. Drijfhout)

TABLE 3. *Bean common mosaic virus* (BCMV) Pathogenicity Groups, Pathogenes, Strains, Differential Cultivars, and Genes for Resistance

Host Resistance Group	Host Differential Cultivar	Host Genes for Resistance	I — P0 — NL1 US1	II — P1 — NL7	III — P2 — NL8	IV — P1.1² — US5	V — P1.2 — NL6 US3 US4	V — P1.2 — US2	V — P1.2 — NL2	VI — P1.1².2 — NL3	VI — P1.1².2 — NL5	VII — P1.1².2² — NL4 US6
Differentials without gene *I*												
1	Dubbele Witte, Sutter Pink, Stringless Green Refugee		+[b]	+	+	+	+	+	+	+	+	+
2	Imuna, Redlands Greenleaf C, Puregold Wax	*bc-u*,[c] *bc-1*	−	+	−	+	+	+	+	+	+	+
3	Redlands Greenleaf B, Great Northern UI 123	*bc-u*, *bc-1²*	−	−	−	+	+	−	−	+	+	+
4	Michelite 62, Sanilac, Red Mexican UI 34	*bc-u*, *bc-2*	−	−	+	−	−	+	+	+	+	−
5	Pinto UI 114	*bc-u*, *bc-1*, *bc-2*	−	−	−	−	−	+	+	+	+	−
6	Great Northern UI 31, Monroe, Red Mexican UI 35	*bc-u*, *bc-1²*, *bc-2²*	−	−	−	−	−	−	−	−	−	+
7	IVT 7214	*bc-u*, *bc-2*, *bc-3*	−	−	−	−	−	−	−	−	−	−
Differentials with gene *I*												
8	Widusa, Black Turtle Soup 1	*I*	−	−	+n	−	±n	−	±n	+n	+n	−
9	Top Crop, Improved Tendergreen, Jubila	*bc-1*, *I*	−	−	−n	−	±n	−	±n	+n	+n	−
10	Amanda	*bc-1²*, *I*	−	−	−n	−	±n	−	−	±n	+n	−
11	IVT 7233	*bc-u*, *bc-1²*, *bc-2²*, *I*	−	−	−n	−	−	−	−	−n	−n	−

[a] Strains US1, US2, US3, US4, US5, and US6 are type, New York 15, Idaho, Western, Florida, and Mexican strains, respectively.

[b] + = Susceptible, systemic mosaic; − = resistant, no systemic symptoms; +n = susceptible, usually all plants with local vein necrosis and systemic necrosis, not temperature dependent; ±n = susceptible, development of systemic necrosis dependent on temperature, the numbers of plants with systemic necrosis increasing with temperature, little or no local necrosis; −n = resistant, no systemic necrosis, pinpoint lesions as local necrosis.

[c] *bc-u* is a recessive, strain-unspecific gene; *bc-1*, *bc-1²*, *bc-2*, *bc-2²*, and *bc-3* are recessive, strain-specific genes; *I* is a dominant, strain-unspecific gene.

10 kb in size. The capsid protein is formed from a single protein subunit with a relative molecular weight 31,000.

Ten major strains (pathotypes) can be differentiated (Table 3). NLI/USI, NL7, US5 (Florida), US2 (NYI5), and NL4/US6 (Mexican) are non-necrosis-inducing BCMV strains. NL2, NL6, US3 (Idaho), and US4 (Western) are temperature-dependent, necrosis-inducing BCMV strains. NL3, NL5, and NL8 are temperature-independent, necrosis-inducing BCMNV strains. Recently, other legume potyviruses, such as *Azuki bean mosaic virus*, *Blackeye cowpea mosaic virus*, and *Peanut stripe virus*, have been classified as strains of BCMV. However, these strains do not usually induce characteristic common mosaic symptoms in *Phaseolus vulgaris* L.

Bean host differentials are divided into two main groups: those with and those without the *I* gene. Eleven resistance groups have been found. Lines IVT 7214 and IVT 7233 are resistant to all 10 strains of the virus. A gene-for-gene relationship has been found between strain-specific resistance genes *bc-1*, *bc-1²*, *bc-2*, and *bc-2²* and members of strains with the same numerical codes.

Disease Cycle and Epidemiology

BCMV has a worldwide distribution because of its high rates (average 35%) of transmission via seeds produced by plants systemically infected prior to bloom; little or no seed transmission occurs if infection occurs after bloom. BCMNV is also seedborne in common bean, but its geographic distribution is more restricted and its incidence considerably lower in most common bean production regions of the world. BCMV

Fig. 120. Winged aphid on a bean leaf. (Courtesy H. F. Schwartz)

predominates in the Western Hemisphere, whereas BCMNV is commonly found in eastern Africa but does occur in the Dominican Republic and Haiti. This virus has been associated with wild legume reservoirs in the African continent. The most important means of secondary spread during the growing season is by aphids (Fig. 120). These viruses can also be transmitted in pollen and are easily transmitted mechanically.

Management

Using certified or virus-free seeds has been shown to reduce bean common mosaic incidence. However, the most satisfactory management method of bean common mosaic is the use of resistant cultivars. Gene combinations of bc-u (strain-nonspecific epistatic gene) plus any of the bc-1, bc-1^2, bc-2, or bc-2^2 genes confer strain-specific, recessive resistance. But since bc-1 and bc-1^2 or bc-2 and bc-2^2 are allelic pairs, it is not possible to have resistance to all strains of the virus in one plant genotype. The combination of bc-u and bc-3 gives recessive resistance to all known strains of BCMV and BCMNV. Likewise, the dominant gene I inhibits all nonnecrotic strains of the virus. However, this gene can be activated by necrosis-inducing strains of BCMV and BCMNV, unless protected by the bc-2^2 or other genes, in which case only restricted necrotic local lesions are produced. Another difficulty with the I gene is that it has a darkening effect on red and yellow seeds. The combination of genes bc-u, bc-2^2, bc-3, and I gives strong double resistance to all known strains. Although it is a considerable breeding task to combine these genes, the combination should result in durable resistance to bean common mosaic.

Selected References

Bos, L. 1971. Bean common mosaic virus. Description of Plant Viruses, No. 73. Commonwealth Mycological Institute and Association of Applied Biologists, Kew, Surrey, England.

Drijfhout, E. 1978. Genetic interaction between *Phaseolus vulgaris* and bean common virus with implications for strain identification and breeding for resistance. Agric. Res. Rep. 872. Centre Agric. Publ. Doc., Wageningen, the Netherlands.

Drijfhout, E., Silbernagel, M. J., and Burke, D. W. 1978. Differentiation of strains of bean common mosaic virus. Neth. J. Plant Pathol. 84:15-26.

Johnson, W. C., Guzman, P., Mandala, D., Mkandawire, A. B. C., Temple, S., Gilbertson, R. L., and Gept, P. 1997. Molecular tagging of the *bc-3* gene for introgression into Andean common bean. Crop Sci. 37:248-254.

Kelly, J. D. 1997. A review of varietal response to bean common mosaic potyvirus in *Phaseolus vulgaris*. Plant Var. Seeds 10:1-6.

Mink, G. I., Vetten, J., Ward, C. W., Berger, P., Morales, F., Myers, J. R., Silbernagel, M. J., and Barnett, O. W. 1994. Taxonomy and classification of legume infecting potyviruses. A proposal from the Potyviridae study group of the plant virus subcommittee of ICTV. Arch. Virol. 139:231-235.

Vandemark, G. J., and Miklas, P. N. 2005. Genotyping common bean for the potyvirus resistance alleles *I* and *bc-1²* with a multiplex real-time polymerase chain reaction assay. Phytopathology 95:499-505.

(Prepared by E. Drijfhout and F. J. Morales;
Revised by F. J. Morales)

Bean Common Mosaic—Black Root

Black root was first observed in 1940 affecting common bean cultivars possessing the 'Corbett Refugee' type of resistance. It is a hypersensitive reaction to infection by certain strains of *Bean common mosaic virus* (BCMV) and *Bean common mosaic necrosis virus* (BCMNV) and appears as a systemic lethal necrosis, producing a brown to black vascular discoloration. This disease may be confused with systemic necroses induced by fungal or bacterial pathogens; but when affected leaves are held against the light, a clear netlike pattern formed by the affected vascular system becomes visible in the case of black root-affected bean plants.

Symptoms

Necrosis in plants possessing monogenic dominant resistance to BCMV and BCMNV, conditioned by the necrosis II gene,

Fig. 121. Pinpoint, necrotic local lesions caused by *Bean common mosaic virus*. (Courtesy F. J. Morales)

Fig. 122. Pinpoint, necrotic local lesions extended to veinal necrosis caused by *Bean common mosaic virus*. (Courtesy F. J. Morales)

Fig. 123. Necrotic "black root" symptoms caused by *Bean common mosaic virus*. (Courtesy H. F. Schwartz)

may start in a mechanically or aphid-inoculated leaf as local lesions (Fig. 121) that extend through the vascular system of the leaf in a starlike fashion (Fig. 122). The vascular necrosis often appears first in young leaves and advances through the vascular system down the stem to the root system. The systemic necrosis spreads to the vascular parts of the stem, causing the wilting and subsequent death of the plant (Fig. 123). Death from systemic necrosis may be rapid or delayed until the pod fill period. When infection occurs late in plant development, pods may show brown to black discoloration in the wall and pod suture (Fig. 124). The type and severity of symptoms depend on host genotype, virus strain, and environmental conditions. Systemic necrosis characteristic of the black root syndrome does not occur in cultivars without the *I* gene or in cultivars with the protected *I* gene.

Causal Agents

The black root syndrome is induced by temperature-dependent, necrosis-inducing BCMV strains NL2, NL6, US3, and US4 and by temperature-independent, necrosis-inducing BCMNV strains NL3, NL5, NL8, and Tanzania 1. However, other potyviruses and particularly some legume potyviruses, such as *Soybean mosaic virus* and *Peanut mottle virus*, can also elicit black root symptoms in *II* gene common bean genotypes.

Disease Cycle and Epidemiology

The necrosis-inducing strains of BCMV and BCMNV are transmitted via the seeds of susceptible common bean genotypes that lack the dominant *I* gene. The planting of dominant *I* gene and recessive *I*⁺ common bean cultivars side by side has caused major epidemics of black root because of the availability of seed-transmitted inoculum and the existence of several aphid species capable of transmitting these viruses in nature. Dominant *I* gene cultivars cannot transmit any BCMV or BCMNV strain via the seeds. Some wild legumes have been shown to act as reservoirs of necrosis-inducing strains of BCMNV in nature.

Management

The most effective management measure for necrosis-inducing strains of BCMV and BCMNV is the production and use of virus-free common bean seeds, specifically of common mosaic-susceptible common bean genotypes. It is equally important to avoid the cocultivation of dominant *I* gene and recessive *I*⁺ gene common bean cultivars in adjacent fields. Genetic resistance to black root can be incorporated in dominant *I* gene genotypes by crossing with sources of recessive genes, such as *bc-2²* and *bc-3*, which protect the *I* gene against necrosis-inducing strains

of BCMV and BCMNV. These protected genotypes show only pinpoint local lesions (*I bc-2²*) or immunity (*I bc-3*).

Selected References

Spence, N., and Walkey, D. G. A. 1994. Bean common mosaic virus and related viruses in Africa. NRI Bull. 63. Natural Resources Institute, Chatham, U.K.

Strausbaugh, C. A., Miklas, P. N., Singh, S. P., Myers, J. R., and Forster, R. L. 2003. Genetic characterization of differential reactions among host group 3 common bean cultivars to NL-3 K strain of *Bean common mosaic necrosis virus*. Phytopathology 93:683-690.

(Prepared by F. J. Morales)

Bean Dwarf Mosaic

Bean dwarf mosaic is an ubiquitous disease of beans in Latin America, where it is recognized under various names, such as chlorotic mottle, mottled dwarf, mosaico anão (dwarf mosaic in Portuguese), and achaparramiento (dwarfing in Spanish). This disease is usually found at a low incidence (<5%) in most countries where it occurs, but major epidemics have affected extensive bean-producing areas of northwestern Argentina, causing total yield losses in susceptible bean genotypes.

The main reservoirs of causal viruses of beans in nature are various species of malvaceous weeds, particularly *Sida* species commonly found in bean-producing regions. The primary symptom induced by the virus in these species consists of a bright yellow mosaic. Other species artificially infected with the causal agent of Sida mosaic include *Datura stramonium* L., *Glycine max* (L.) Merr., *Lycopersicon esculentum* Mill., *Nicandra physalodes* (L.) Gaertn., *Nicotiana tabacum* L., and *Solanum tuberosum* L.

Symptoms

Bean plants affected by bean dwarf mosaic at the cotyledonary leaf stage generally exhibit marked stunting (Fig. 125) as shortening of internodes, foliar malformation, flower abortion, and pod distortion. Plants affected at a later stage show irregular mottling or mosaic symptoms. Chlorotic areas of variegated leaves grow more slowly than do the adjacent tissues, which results in noticeable leaf malformation. Chlorotic lesions may develop into yellow patches that are irregularly distributed in the foliage (Fig. 126). Flower abortion is commonly observed in systemically affected plants, but if pods are formed, they become severely distorted (Fig. 127) and do not produce normal seeds. Highly susceptible genotypes infected soon after

Fig. 124. Necrosis of pod walls and pod sutures (right half) caused by *Bean common mosaic virus*. (Courtesy H. F. Schwartz)

Fig. 125. Plant stunting caused by *Bean dwarf mosaic virus*. (Courtesy F. J. Morales)

germination do not develop past the cotyledonary leaf stage, because all the internodes become shortened. This gives the plant a bushy appearance.

Causal Agent

Bean dwarf mosaic virus (BDMV) is a species of the genus *Begomovirus* of the family *Geminiviridae*, and it has a bipartite genome. Virions characteristically consist of two incomplete, icosahedral particles, each one approximately 20 nm in diameter. These geminate virions contain a single molecule of single-stranded DNA 2.6 kb in size. BDMV is closely related to *Abutilon mosaic virus*. BDMV induces the formation of viruslike particle aggregates in the nucleus of infected bean cells.

Disease Cycle and Epidemiology

BDMV is transmitted by mechanical means with some difficulty. In nature, the virus is transmitted by the whitefly *Bemisia tabaci* (Gennadius) in a persistent manner, but no evidence has been found of virus multiplication in the vector or of transovarial transmission. BDMV does not seem to be seed-transmitted in either common bean or *Sida* spp. Epidemics of BDMV are associated with the occurrence of large populations of the whitefly vector at the beginning of the bean planting season, which, in turn, are favored by the presence of suitable whitefly reproductive hosts, insecticide overuse, and warm/dry conditions in the affected regions.

Management

Avoidance of host plants favored by *B. tabaci* is highly recommended, especially cotton, soybean, tobacco, tomato, potato, and eggplant. Resistant cultivars are available and should be used whenever possible for production and as parents in breeding programs. Some of the more resistant cultivars include Porrillo Sintetico, DOR 41, Red Mexican UI 35, and pinto UI 114. The best sources of resistance to *Bean golden mosaic virus* are also effective against BDMV. The vector can be managed with systemic insecticides, but chemical management is very expensive and hazardous.

Selected References

Costa, A. S. 1975. Increase in the populational density of *Bemisia tabaci*, a threat of widespread virus infection of legume crops in Brazil. Pages 27-49 in: Tropical Diseases of Legumes. J. Bird and K. Maramorosch, eds. Academic Press, New York.

Galvez, G. E., and Morales, F. J. 1989. Whitefly-transmitted viruses. Pages 379-406 in: Bean Production Problems in the Tropics, 2nd ed. H. F. Schwartz and M. A. Pastor-Corrales, eds. Centro Internacional de Agricultura Tropical (CIAT), Cali, Colombia.

Hidayat, S. H., Gilbertson, R. L., Hanson, S. F., Morales, F. J., Ahlquist, P., Russell, D. R., and Maxwell, D. P. 1993. Complete nucleotide sequences of the infectious cloned DNAs of bean dwarf mosaic geminivirus. Phytopathology 83:181-187.

Morales, F. J., and Niessen, A. I. 1988. Comparative responses of selected *Phaseolus vulgaris* germ plasm inoculated artificially and naturally with bean golden mosaic virus. Plant Dis. 72:1020-1023.

Morales, F., Niessen, A., Ramírez, B., and Castaño, M. 1990. Isolation and partial characterization of a geminivirus causing bean dwarf mosaic. Phytopathology 80:96-101.

(Prepared by F. J. Morales)

Fig. 126. Foliar symptoms of bean dwarf mosaic. (Courtesy F. J. Morales)

Fig. 127. Distortion of bean pods caused by *Bean dwarf mosaic virus*. (Courtesy F. J. Morales)

Bean Golden Mosaic

Bean golden mosaic was first described during the early 1960s in São Paulo, Brazil, as a disease of minor importance. However, this disease eventually reached epidemic proportions throughout the main bean-growing regions of Brazil, Argentina, and the lowlands of Bolivia. Bean golden mosaic is considered the main biotic constraint of common bean production in this region of South America, particularly during the dry, warmer months of the year. Yield losses are related to the time and frequency of virus inoculation and the common bean genotype affected. If infection occurs within the first 2 weeks after plant emergence, yield losses may be total in susceptible cultivars and up to 40% in disease-resistant common bean cultivars.

Symptoms

Bean golden mosaic causes intense systemic yellowing in most common bean landraces grown in South America, particularly in common bean genotypes of Andean (Nueva Granada race) origin (Fig. 128). In black-seeded, Mesoamerican common bean genotypes or cultivars derived from these sources, a less-intense yellowing and systemic mosaic symptoms are observed (Fig. 129). Systemically infected plants usually abort their flowers or produce distorted pods (Fig. 130) bearing few seeds that are small and damaged. Flower abortion is aggravated by high temperatures. All known common bean genotypes are susceptible to the pathogen, but genotypes differ in their level of susceptibility/resistance.

The causal virus of bean golden mosaic has a narrow host range. The virus infects *Macroptilium lathyroides* (L.) Urb.,

M. longipedunculatum (Mart. ex Benth.) Urb., *Phaseolus lunatus* L., and *Phaseolus vulgaris* L., but not *Cajanus cajun* (L.) Millsp., *Cassia occidentalis* L., or *Rhynchosia minima* (L.) DC.

Causal Agent

Bean golden mosaic virus (BGMV) is a species of the genus *Begomovirus* of the family *Geminiviridae*. BGMV has geminate, near-icosahedral (18 × 30 nm) particles containing two circular single-stranded DNA molecules 2,617 and 2,580 nucleotides in size. BGMV is not manually transmissible, but its infective DNA can be mechanically inoculated by biolistic methods. Viroplasms found in the nucleus are often observed as fibrillar rings.

Disease Cycle and Epidemiology

BGMV is transmitted by different biotypes of the whitefly *Bemisia tabaci* (Gennadius). One of these biotypes (B) is sometimes referred to as *B. argentifolii* Bellows & Perring. Adult *B. tabaci* may acquire the virus from systemically infected common bean plants in a few minutes, but longer virus acquisition feeding times are required under experimental conditions to achieve high virus transmission rates. Viruliferous whiteflies can transmit the virus for periods ranging from a few days to life. BGMV does not seem to either propagate or be transovarially transmitted in *B. tabaci*.

Bean golden mosaic epidemics depend on the presence of a suitable reproductive host for the whitefly vector and on the dry and warm environmental conditions that favor whitefly reproduction and vector movement. The traditional bean plantings during the seca (dry) and warm period at the beginning of the year (January–March) had to be abandoned because of the high incidence of BGMV. Pesticide abuse leading to insecticide-insensitive vectors is another factor that has been consistently associated with BGMV epidemics.

Management

BGMV incidence is usually higher when common bean is planted with the first rains or irrigation following a dry, warm period, which favors the buildup of *B. tabaci* populations. Crop damage is more severe under these conditions when a good reproductive host for the whitefly is planted before common bean and when harvest (physiological maturity) of the whitefly host coincides with planting of the bean crop. This phenomenon is explained by the migration of whiteflies from the reproductive host to common bean plantings. In South America, its incidence is higher in common bean seeded in the first quarter of the year following the harvest of soybean, the main reproductive host of *B. tabaci* in this region.

All common bean genotypes are more susceptible within the first 2 weeks of vegetative growth. Hence, only systemic

insecticides applied at sowing may be effective in reducing BGMV damage. Cultural practices designed to avoid planting beans in the presence of moderate to large populations of *B. tabaci* should effectively complement other disease management measures. Genetic resistance is the most critical component of any integrated pest management strategy implemented. Black-seeded common bean cultivars have shown adequate levels of BGMV resistance in South America, but other bean genotypes of Andean or Mexican origin also possess complementary mechanisms of BGMV resistance. Despite the difficulties involved, transgenic common bean genotypes possessing moderate levels of BGMV resistance have already been obtained in Brazil.

Selected References

Costa, A. S. 1965. Three whitefly-transmitted virus diseases of beans in Sao Paulo, Brazil. FAO Plant Prot. Bull. 13:121-130.

Costa, A. S. 1978. Whitefly-transmitted plant diseases. Annu. Rev. Phytopathol. 14:429-449.

Galvez, G. E., and Morales, F. J. 1989. Whitefly-transmitted viruses. Pages 379-406 in: Bean Production Problems in the Tropics, 2nd ed. H. F. Schwartz and M. A. Pastor-Corrales, eds. Centro Internacional de Agricultura Tropical (CIAT), Cali, Colombia.

Gilbertson, R. L., Faria, J. C., Ahlquist, P., and Maxwell, D. P. 1993. Genetic diversity in geminiviruses causing bean golden mosaic disease: The nucleotide sequence of the infectious cloned DNA components of a Brazilian isolate of bean golden mosaic geminivirus. Phytopathology 83:709-715.

Goodman, R. M. 1981. Geminiviruses. Pages 879-910 in: Handbook

Fig. 129. Bright golden mosaic caused by *Bean golden mosaic virus*. (Courtesy F. J. Morales)

Fig. 128. Foliar symptoms of bean golden mosaic. (Courtesy F. J. Morales)

Fig. 130. Pod distortion caused by *Bean golden mosaic virus*. (Courtesy F. J. Morales)

of Plant Virus Infections: Comparative Diagnosis. E. Kurstak, ed. Elsevier/North Holland Biomedical Press, Amsterdam.

Morales, F. J., and Anderson, P. K. 2000. The emergence and dissemination of whitefly-transmitted geminiviruses in Latin America. Arch. Virol. 146:415-441.

Morales, F. J., and Niessen, A. I. 1988. Comparative responses of selected *Phaseolus vulgaris* germ plasm inoculated artificially and naturally with bean golden mosaic virus. Plant Dis. 72:1020-1023.

(Prepared by F. J. Morales)

Bean Golden Yellow Mosaic

Bean golden yellow mosaic resembles bean golden mosaic, and both were considered to be the same disease for more than 2 decades. However, these diseases are caused by related but different virus species. Bean golden yellow mosaic was first observed in the late 1960s, and it spread mainly in Central America, southern Mexico, and the Caribbean. Yield losses depend on the time of infection and the common bean genotype affected. Susceptible common bean genotypes infected at the seedling stage usually sustain 100% yield losses.

Symptoms

Symptoms associated with bean golden yellow mosaic vary considerably depending on the virus isolate, common bean genotype, time of infection, and environmental conditions. Susceptible bean genotypes infected soon after emergence under warm and dry conditions generally display an intense systemic yellowing (Fig. 131). Plant malformation is more apparent in common bean genotypes of Mesoamerican origin, but some Andean genotypes have dwarfing genes that are sometimes expressed under natural conditions when plants are infected at the seedling stage. Black-seeded and some virus-resistant genotypes usually display partially systemic symptoms of moderate intensity (Fig. 132). Flower abortion is common in infected Mesoamerican common bean genotypes, particularly at high temperatures. *Phaseolus lunatus* L., a common host of the virus, shows intense foliar and vein yellowing but little dwarfing.

Causal Agent

Bean golden yellow mosaic virus (BGYMV) is a species of the genus *Begomovirus* of the family *Geminiviridae*. Virions are geminate, 18–20 nm in diameter, and 30 nm in length for the dimer. The BGYMV isolates characterized so far have two molecules of single-stranded DNA approximately 2,640 and 2,600 nucleotides in size, and a coat protein subunit with a relative molecular weight of 27,000. Characteristic ring-shaped viroplasms are observed in the nuclei of infected plant cells.

Disease Cycle and Epidemiology

BGYMV is transmitted by at least two biotypes of the whitefly *Bemisia tabaci* (Gennadius). Adult *B. tabaci* may acquire the virus from systemically infected bean plants in a few minutes, but longer virus acquisition feeding times are required under experimental conditions before whiteflies achieve high virus transmission rates. Viruliferous whiteflies can transmit the virus for periods ranging from a few days to life. BGYMV has not been shown to either propagate or be transovarially transmitted in *B. tabaci*.

Bean golden yellow mosaic epidemics depend on the presence of suitable reproductive hosts for the whitefly vector and on the dry and warm conditions that favor whitefly reproduction and vector movement. These conditions are often found during the dry period that extends from December through April in Mesoamerica and the Caribbean, where a number of horticultural and industrial crops, such as tomato, eggplant, broccoli, soybean, and tobacco, may act as reproductive hosts for *B. tabaci*. Moreover, high-value horticultural and industrial crops are constantly treated with a mixture of insecticides that promotes pesticide-resistant whitefly populations.

Management

Genetic resistance has been the most effective BGYMV management method used to date, although all known common bean cultivars are susceptible. The first sources of resistance identified were black-seeded, Mesoamerican bean genotypes that were, at best, tolerant and had the ability to escape infection under field conditions. The resistance in these genotypes can be transferred to susceptible bean cultivars but is not durable. Other BGYMV-resistant genes in *Phaseolus vulgaris* L.

Fig. 131. Systemic yellowing caused by *Bean golden yellow mosaic virus*. (Courtesy F. J. Morales)

Fig. 132. Partially systemic symptoms of bean golden yellow mosaic. (Courtesy F. J. Morales)

races from different gene pools, such as the *bgm-1* (also represented as *bgm*) gene in genotypes of the Mexican Durango race or the *bgm-2* gene of Mesoamerican or Andean origin, are known. Systemic insecticide applications have been shown to be effective when applied at planting time, but *B. tabaci* has developed resistance to most of the systemic and contact insecticides used in the past. Recently developed pesticides, such as imidacloprid, are more effective in managing *B. tabaci*. In the Dominican Republic, a whitefly host-free period before planting one (November) rather than two (September and January) crops of bean has significantly reduced the disease. Cultivation of common bean only during the rainy seasons, when whitefly population densities are generally low, has been another effective virus management practice.

Selected References

Faria, J. C., Gilbertson, R. L., Hanson, S. F., Morales, F. J., Ahlquist, P., Loniello, A. O., and Maxwell, D. P. 1994. Bean golden mosaic geminivirus type II isolates from the Dominican Republic and Guatemala: Nucleotide sequences, infectious pseudorecombinants, and phylogenetic relationships. Phytopathology 84:321-329.

Morales, F. J., and Anderson, P. K. 2000. The emergence and dissemination of whitefly-transmitted geminiviruses in Latin America. Arch. Virol. 146:415-441.

Morales, F. J., and Singh, S. P. 1991. Genetics of resistance to bean golden mosaic virus in *Phaseolus vulgaris* L. Euphytica 52:113-117.

Morales, F. J., and Singh, S. P. 1993. Breeding for resistance to bean golden mosaic virus in an interracial population of *Phaseolus vulgaris* L. Euphytica 67:59-63.

Singh, S. P., Morales, F. J., Miklas, P. N., and Terán, H. 2000. Selection for bean golden mosaic resistance in intra- and interracial bean populations. Crop Sci. 40:1565-1572.

Velez, J. J., Basset, M. J., Beaver, J. S., and Molina, A. 1998. Inheritance of resistance to bean golden mosaic virus in common bean. J. Am. Soc. Hortic. Sci. 123:628-631.

(Prepared by F. J. Morales)

Bean Mild Mosaic

Bean mild mosaic is a highly contagious but mild disease of common bean, usually overlooked under field and experimental conditions. The causal virus is often detected in mixed infections with other viruses inducing synergistic infections observed as witches'-broom or severe leaf malformation. Bean mild mosaic has been isolated from common bean plants grown under natural conditions in El Salvador, Colombia, and the United States. Experimental infection of common bean plants has occurred in Europe. All common bean genotypes inoculated to date are susceptible. *Phaseolus acutifolius* A. Gray and *P. coccineus* L. are also susceptible, but *P. lunatus* L. possesses genes for resistance to bean mild mosaic.

Symptoms

Bean mild mosaic symptoms usually start as mild vein yellowing of the youngest trifoliolate leaves. Symptoms tend to disappear with plant age, and only mild chlorosis or mosaic symptoms remain in mature plants (Fig. 133). The causal virus infects other legumes such as *Glycine max* (L.) Merr., *Lablab purpureus* (L.) Sweet, *Macroptilium atropurpureum* (Moc. & Sessé ex DC.) Urb., *M. lathyroides* (L.) Urb., *Rhynchosia minima* (L.) DC., and *Sesbania exaltata* (Raf.) Rydb. ex A. W. Hill. No infection has been observed following inoculation of several nonleguminous species, including grasses, solanaceous species (including *Nicotiana benthamiana* Domin.), crucifers, cucurbits, and several annual ornamental crops.

Causal Agent

Bean mild mosaic virus (BMMV) is a species of the genus *Carmovirus* of the family *Tombusviridae*. BMMV particles are isometric, approximately 28 nm in diameter, and contain single-stranded RNA of molecular weight of 1.27×10^6, which composes 20% of the particle weight. Particles sediment as a single component with an $s_{20,w}$ of about 127 S. The capsid protein subunit of BMMV has a relative molecular weight of approximately 40,000 and the single-stranded RNA is 4.2 kb in size.

BMMV is serologically unrelated to 35 other viruses with spherical particles, including 10 usually associated with legumes and 11 other viruses with single sedimenting components. Virus particle aggregates can be observed in the cytoplasm and vacuoles of infected bean cells.

Disease Cycle and Epidemiology

The virus is readily sap-transmitted with or without buffer or abrasives. It is also transmitted by several species of beetles, including the Mexican bean beetle (*Epilachna varivestis* Mulsant), spotted cucumber beetle (*Diabrotica undecimpunctata howardi* Barber), banded cucumber beetle (*D. balteata* LeConte), bean leaf beetle (*Cerotoma ruficornis* (Olivier)), and *Gynandrobrotica variabilis* Jacoby. The virus was not transmitted by two other beetle species (*D. adelpha* Harold and *Paranapiacaba waterhousei* Jac.), leaf miners (*Liriomyza munda* Frick), or greenhouse whiteflies (*Trialeurodes vaporariorum* (Westwood)). BMMV also spreads via root contact in the soil. It can be soil-transmitted in the absence of plant-parasitic nematodes. Unless extreme caution is taken, BMMV can spread uncontrollably in the greenhouse among beans. The virus can be seedborne in common bean.

Management

The mild symptoms and rare occurrence of bean mild mosaic has not prompted the adoption of any specific virus management measures. Immunity to BMMV has been found in *Phaseolus anisotrichus* Schlechtend., *P. filiformis* Benth., *P. leptostachyus* Benth., and *P. lunatus* but has not been detected in common bean. Management of chrysomelid vectors and regular serological assays of common bean plants grown under experimental or field conditions should reduce the incidence of this disease.

Selected References

Hobbs, H. A. 1981. Transmission of bean curly dwarf mosaic virus and bean mild mosaic virus by beetles in Costa Rica. Plant Dis. 65:491-492.

Jayasinghe, W. U. 1982. Chlorotic mottle of bean (*Phaseolus vulgaris*

Fig. 133. Leaf symptoms of bean mild mosaic. (Courtesy F. J. Morales)

L.). CIAT Ser. 09EB. Centro Internacional de Agricultura Tropical (CIAT), Cali, Colombia.

Meiners, J. P., Waterworth, H. E., Lawson, R. H., and Smith, F. F. 1977. Curly dwarf mosaic disease of beans from El Salvador. Phytopathology 67:163-168.

Morales, F. J., and Gámez, R. 1989. Beetle-transmitted viruses. Pages 363-377 in: Bean Production Problems in the Tropics, 2nd ed. H. F. Schwartz and M. A. Pastor-Corrales, eds. Centro Internacional de Agricultura Tropical (CIAT), Cali, Colombia.

Sepulveda, P. S., Saettler, A. M., Gillet, J., and Ramsdell, D. C. 1993. Identification of bean mild mosaic virus as a new virus of field grown beans in the United States. Fitopatologia 28:63-69.

Waterworth, H. E., Meiners, J. P., Lawson, R. H., and Smith, F. F. 1977. Purification and properties of a virus from El Salvador that causes mild mosaic in bean cultivars. Phytopathology 67:169-173.

(Prepared by H. E. Waterworth; Revised by F. J. Morales)

Bean Necrosis Mosaic

Bean necrosis mosaic was described as a disease of common bean in Brazil around 1957. The incidence of this disease in the state of São Paulo is sporadic and ranges between 5 and 20%.

Symptoms

Common bean plants affected by this disease show simultaneous mosaic and necrosis symptoms (Fig. 134). The mosaic symptoms are mild, and the necrosis is often expressed as irregular or concentric ring spots. Diseased common bean plants do not show striking malformation or dwarfing symptoms. Mechanically inoculated plants may exhibit local lesions.

Causal Agent

Bean necrosis mosaic is caused by strains of *Tomato spotted wilt virus* (TSWV) adapted to common bean. When TSWV strains adapted to solanaceous species infect common bean, the disease is called vira cabeça (turning heads) in Portuguese. TSWV is a species of the genus *Tospovirus* of the family *Bunyaviridae*. Virions are enveloped, 85 nm in diameter (round), and contain three genome segments of single-stranded RNA approximately 8.9 (large), 5.4 (medium), and 2.9 (small) kb in size.

Disease Cycle and Epidemiology

TSWV is not transmitted through seeds or mechanically under natural conditions. Transmission of TSWV in the field is dependent on the presence of thrips vectors, such as *Thrips* and *Frankliniella* species. Transmission of the virus strains adapted to common bean in Brazil occurs within the field. Other TSWV strains adapted to other hosts are usually transmitted to common bean from other virus reservoirs and not from bean to bean.

Management

There are sources of resistance in *Phaseolus vulgaris* L. to TSWV. Chemical management of the insect vector is possible, particularly in the case of the common bean strains of TSWV, which are disseminated inside common bean fields.

Selected References

Costa, A. S. 1987. Fitoviroses do feijoeiro no Brasil. Pages 173-256 in: Feijão: Fatores de Produção e Qualidade. E. A. Bulisani, ed. Fundação Cargill, Campinas, São Paulo, Brazil.

Gibbs, A. J. 1996. Tomato spotted wilt tospovirus. Pages 1312-1315 in: Viruses of Plants. A. A. Brunt, K. Crabtree, M. J. Dallwitz, A. J. Gibbs, and L. Watson, eds. CAB International, Wallingford, U.K.

(Prepared by F. J. Morales)

Bean Pod Mottle

Bean pod mottle was first described in 1948 as affecting common bean in South Carolina. This disease affects common bean only in the southeastern and midwestern United States. The virus is rarely found in beans but occurs commonly in soybean and the perennial weed *Desmodium* spp.

Symptoms

Systemically infected bean leaves exhibit persistent symptoms ranging from severe to mild mosaic with malformation (Fig. 135). The pods of many susceptible cultivars are malformed and mottled and may contain poorly developed seeds. Common bean cultivars susceptible to systemic infection include Black Valentine, Black Wax, Burpee's Stringless Green Pod, Cherokee Wax, Dwarf Horticultural, Pencil Pod, Red Kidney, and Tendergreen. Local lesions are formed on inoculated leaves of Blue Lake, Bountiful, Keeney Refugee, Michelite, Scotia, Top Crop, and certain navy and pinto cultivars. The local lesions on pinto bean are characteristic and suitable for virus assay. These lesions appear 3–4 days after inoculation and are light brown, diffuse, and circular.

The host range is limited to legumes, including soybean, *Desmodium* spp., a *Lespedeza* sp., *Stizolobium deeringianum* Bort, *Trifolium incarnatum* L., and some cultivars of *Vigna unguiculata* (L.) Walp.

Fig. 134. Leaf symptoms of bean necrosis mosaic. (Courtesy F. J. Morales)

Fig. 135. Leaves and pods infected with *Bean pod mottle virus*. (Courtesy H. A. Scott)

Causal Agent

Bean pod mottle is caused by *Bean pod mottle virus* (BPMV), a species of the genus *Comovirus*. The virus is icosahedral, about 28 nm in diameter, and exhibits three centrifugal components: top (capsid protein), middle (4.8 kb of single-stranded RNA), and bottom (8 kb of single-stranded RNA). It is a divided genome virus in which the middle and bottom components are required for infection. BPMV is easily detected in gel diffusion tests with crude sap.

Disease Cycle and Epidemiology

Since BPMV is easily sap transmissible, some spread of the virus in field-grown beans probably occurs by plant-to-plant contact. However, leaf-feeding beetles (order Coleoptera) belonging to the families *Chrysomelidae*, *Coccinellidae*, and *Meloidae* are undoubtedly the chief means of virus transmission. Transmission of BPMV from infected soybean or bean to bean has been demonstrated only with *Ceratoma trifurcata* (Forster), *Diabrotica undecimpunctata howardi* Barber, and *Epilachna varivestis* Mulsant.

The bean leaf beetle (*C. trifurcata*) transmits BPMV at high levels of efficiency for several days and is the most important vector in the field. *D. undecimpunctata howardi* and *E. varivestis* are not as efficient as the bean leaf beetle. *E. varivestis* rarely transmits the virus more than 1 day beyond the acquisition feeding.

Management

BPMV is of limited economic importance in common bean, probably because of natural resistance in *Phaseolus vulgaris* L. genotypes. The virus is not seedborne in beans. Broadleaf weeds adjacent to bean fields should be managed. The perennial weeds *Desmodium paniculatum* (L.) DC. and *Desmodium canadense* (L.) DC. may be naturally infected with BPMV and, since bean leaf beetles feed on these plants, they have been implicated as reservoirs of virus. Beans should be separated from other leguminous crops with tall crops, such as corn, to restrict entry of beetles. Bean leaf beetles overwintering as adults in field trash and along the edges of woods remain viruliferous for several months during their dormant phase and transmit BPMV when they become active in the spring.

Strategies for managing BPMV have been directed toward the beetle vector but have not been generally applied or critically evaluated. Rows of soybean planted along field edges 10–14 days before the regular planting attract and concentrate overwintering beetles. Insecticide applied after the colonizing (overwintering) beetles aggregate can decrease the number of potential beetle vectors.

Selected References

Ghabrial, A. A., and Schultz, F. J. 1983. Serological detection of bean pod mottle virus in bean leaf beetles. Phytopathology 73:480-483.

Lin, M. T., and Hill, J. H. 1983. Bean pod mottle virus: Occurrence in Nebraska and seed transmission in soybeans. Plant Dis. 67:230-233.

Skotland, C. B. 1958. Bean pod mottle virus of soybeans. Plant Dis. Rep. 42:1155-1156.

Walters, H. J. 1964. Transmission of bean pod mottle virus by bean leaf beetles. Phytopathology 54:240.

(Prepared by H. A. Scott; Revised by F. J. Morales)

Bean Rugose Mosaic

Bean rugose mosaic was first described as a severe disease of common bean in Costa Rica, and later in El Salvador and Guatemala. A disease known as mosaico-em-desenhos in Brazil was shown to be caused by a closely related virus. Bean rugose mosaic was considered a major threat to common bean production in the late 1970s and early 1980s, but this disease has practically disappeared since then.

Symptoms

As its name implies, bean rugose mosaic usually induces severe foliar rugosity and pod malformation (Fig. 136). Common bean reactions to the virus include systemic infection, local lesions, and immunity. The causal virus is apparently restricted to legumes, such as *Cicer arietinum* L., *Glycine max* (L.) Merr., *Macroptilium lathyroides* (L.) Urb., *Phaseolus acutifolius* A. Gray, *P. lunatus* L., *P. vulgaris* L., *Pisum sativum* L., *Trifolium incarnatum* L., *Vicia faba* L., and *Vigna unguiculata* (L.) Walp. Common bean cultivars Kentucky Wonder, Michelite, Stringless Green Refugee, and Tendergreen are immune. Cultivars Jamapa, Sanilac, Top Crop, and some pinto cultivars react with local lesions when inoculated mechanically.

Causal Agent

Bean rugose mosaic virus (BRMV) is a species of the genus *Comovirus* of the family *Comoviridae*. Virions are isometric and 28 nm in diameter. The comoviruses are characterized by three centrifugal components: top (devoid of RNA), middle, and bottom, with infectivity dependent upon the presence of both middle and bottom components. Virus preparations contain two RNA species with molecular weights of 2.3×10^6 and 1.4×10^6 and two polypeptides with molecular weights of 38,000 and 22,000. BRMV strains are serologically identical or closely related, and as one of the five comovirus serogroups, they are readily differentiated in gel diffusion serological tests.

Disease Cycle and Epidemiology

BRMV is transmitted by leaf-feeding beetles, but the length of time the beetle transmits and its transmission efficiency depend upon the beetle species. *Cerotoma ruficornis* (Olivier) transmits at a higher rate on the first day (80%) than does *Diabrotica balteata* LeConte or *D. adelpha* Harold (20 and 10%, respectively). *C. ruficornis* transmits the virus for up to 10 days after acquisition feeding. *Diabrotica* spp. transmit for only 2–3 days.

Management

Resistance to BRMV is available in common bean cultivars grown in Central and South America. BRMV resistance is governed by three alleles, the first of which is dominant over the other two and confers immunity to the virus. The second is dominant over the third and confers hypersensitivity, and the third conditions susceptibility (systemic infection) to BRMV.

Fig. 136. Plant infected with *Bean rugose mosaic virus*. (Courtesy F. J. Morales)

In Central America, corn/bean polycultures contribute to the occurrence of larger populations of *D. balteata* than do bean monocultures. If an intercrop contains at least one nonhost plant for a given beetle species, the number of beetles in the intercrop is reduced relative to the number in the monocrop. Chemical management of chrysomelid vectors is feasible.

Selected References

Altieri, M. A., Francis, C. A., Van Schoonhoven, A., and Doll, J. D. 1978. A review of insect prevalence in maize (*Zea mays* L.) and bean (*Phaseolus vulgaris* L.) polycultural systems. Field Crops Res. 1:33-49.

Galvez, G. E., Cardenas-Alonso, M., Kitajima, E. W., Diaz-Ch., A. J., and Nieto-C., M. P. 1977. Purification, serology, electron microscopy and properties of the ampollado strain of bean rugose mosaic virus. Turrialba 27:343-350.

Gámez, R. 1972. Los virus del frijol en Centroamerica, II: Algunas propiedades y transmision por crisomelidos del virus del mosaico rugoso del frijol. Turrialba 22:249-257.

Morales, F. J., and Gámez, R. 1989. Beetle-transmitted viruses. Pages 363-377 in: Bean Production Problems in the Tropics. H. F. Schwartz and M. A. Pastor-Corrales, eds. Centro Internacional de Agricultura Tropical (CIAT), Cali, Colombia.

Risch, S. J. 1989. Fewer beetle pests on beans and cowpeas interplanted with banana in Costa Rica. Turrialba 30:229-230.

Risch, S. J. 1989. The population dynamics of several herbivorous beetles in a tropical agroecosystem: The effect of intercropping corn, beans and squash in Costa Rica. J. Appl. Ecol. 17:593-612.

(Prepared by H. A. Scott; Revised by F. J. Morales)

Bean Severe Mosaic

Bean severe mosaic was previously referred to as bean curly dwarf mosaic. This disease was first observed in El Salvador in 1971, and it is now present in Guatemala and Honduras. One of the causal comoviruses also infects common bean cultivars under natural conditions in Brazil and Venezuela. Yields may be greatly reduced and total losses can occur, particularly in common bean cultivars that develop top or systemic necrosis.

Symptoms

Most common bean landraces show mosaic and varying degrees of leaf malformation, mainly rugosity and curling (Fig. 137). Dwarfing and proliferation may be observed in mixed infections with other viruses, such as *Bean common mosaic virus*, *Bean mild mosaic virus*, or *Southern bean mosaic virus*.

Fig. 137. Leaf malformation caused by *Quail pea mosaic virus*. (Courtesy F. J. Morales)

The most severe symptom observed in some common bean genotypes infected by severe-mosaic-inducing viruses is top necrosis (Fig. 138), which results in plant death. Restricted or systemic necrosis (Fig. 139) occurs in common bean cultivars that possess monogenic dominant resistance to *Bean common mosaic virus*. The development of necrosis, however, is influenced by environmental factors. Mechanically inoculated plants may exhibit local lesions in the inoculated leaves. The causal viruses can be found in the cytoplasm of infected epidermal or parenchymal cells of more than 50 bean genotypes and 16 species of legumes, forming large crystalline inclusion bodies.

Causal Agents

Bean severe mosaic was originally described as a disease caused by a strain of *Quail pea mosaic virus* (QPMV). However, further research has shown that this disease can be induced by different comoviruses related to QPMV, *Cowpea severe mosaic virus* (CPSMV), or both. These viruses belong to the genus *Comovirus* of the family *Comoviridae*, and, consequently, they have three types of isometric particles that are 25 nm in diameter (Fig. 140), one empty and the other two containing single-stranded RNA molecules 6.0 (RNA1) and 3.7 (RNA2) kb in size. Three capsid protein subunits with relative molecular weights of 39,000, 22,000, and 4,000 are usually associated with these viruses. Comoviruses are relatively good immunogens, and specific antisera have been obtained. QPMV is also closely serologically related to the CPSMVs, but it has a more restricted pathogenic spectrum in common bean. QPMV has not been detected outside of Latin America,

Fig. 138. Top necrosis caused by *Quail pea mosaic virus*. (Courtesy F. J. Morales)

Fig. 139. Systemic necrosis caused by *Quail pea mosaic virus*. (Courtesy F. J. Morales)

whereas CPSMV is widely distributed in this region. Research is in progress to determine the identity of the causal agents of bean severe mosaic.

Disease Cycle and Epidemiology

Comoviruses are readily sap-transmitted, with or without the use of abrasives or buffers, but they have not been observed to be seedborne in common bean. Legume comoviruses are transmitted by several species of chrysomelid beetles, such as the spotted and banded cucumber beetles (*Diabrotica undecimpunctata howardi* Barber and *D. balteata* LeConte, respectively), the Mexican bean beetle (*Epilachna varivestis* Mulsant), the flea beetle (*Cerotoma ruficornis* (Olivier)), and species of the genera *Gynandrobrotica* and *Paranapiacaba*. These chrysomelid beetles retain the virus for a few days following virus acquisition from susceptible plant species, mainly weeds growing in the vicinity of bean plantings.

Management

Although none of the bean genotypes tested so far have shown immunity to any of the comoviruses associated with bean severe mosaic, there are some bean genotypes that are not appreciably affected by the causal viruses. These bean genotypes have been shown to possess recessive genes (*anv* and *lnv*) for the apical and localized necroses reactions, respectively, induced by comoviruses that cause bean severe mosaic in some common bean genotypes. Two independently inherited dominant genes, *Anv* and *Lnv*, are responsible for the apical necrosis and characteristic localized necrosis, respectively, observed in some susceptible bean genotypes.

The chemical management of chrysomelid beetle vectors usually reduces the incidence of comoviruses that cause bean severe mosaic. Corn, a crop frequently grown in association with common bean in Latin America, is a well-known host of some chrysomelid vectors and, consequently, bean severe mosaic is more frequently observed under this mixed cropping system.

Selected References

Debrot, E. A., and Centeno, F. 1986. Occurrence of quail pea mosaic virus (QPMV) infecting beans and soybeans in Venezuela. (Abstr.) Phytopathology 76:1089-1090.

Hobbs, H. A. 1981. Transmission of bean curly dwarf mosaic virus and bean mild mosaic virus by beetles in Costa Rica. Plant Dis. 65:491-492.

Meiners, J. P., Waterworth, H. E., Lawson, R. H., and Smith, F. F. 1977. Curly dwarf mosaic disease of beans from El Salvador. Phytopathology 67:163-168.

Moore, B. J. 1973. Quail pea mosaic virus: A new member of the Comovirus group. Plant Dis. Rep. 57:311-315.

Morales, F. J., and Castaño, M. 1992. Increased disease severity induced by some comoviruses in bean genotypes possessing monogenic dominant resistance to bean common mosaic potyvirus. Plant Dis. 76:570-573.

Morales, F. J., and Singh, S. P. 1997. Inheritance of the mosaic and necroses reactions induced by comoviruses in *Phaseolus vulgaris* L. Euphytica 93:223-226.

(Prepared by F. J. Morales)

Bean Southern Mosaic

This disease was first observed in 1941, affecting common bean in Louisiana, a southern state; hence, the misnomer of southern bean mosaic. This disease is probably distributed worldwide but its characteristic mild symptoms are often overlooked or confused with other biotic or abiotic problems of common bean. The severity of this disease is often increased considerably in mixed infections with other common bean viruses. The disease may be particularly severe in other *Phaseolus* species, such as *P. acutifolius* A. Gray.

Symptoms

Bean southern mosaic symptoms on common bean are usually mild, consisting of a lighter olive green color of the infected foliage (Fig. 141). Rugosing, stunting, and witches'-broom symptoms are often associated with mixed infections or interspecific hybrids with *P. acutifolius* (Fig. 142).

Symptoms on pods may appear as dark green, irregularly shaped, water-soaked, blotched areas on green-podded types and as greenish yellow areas on yellow-podded types. Pod deformation with poor seed set and reduced seed size may also occur.

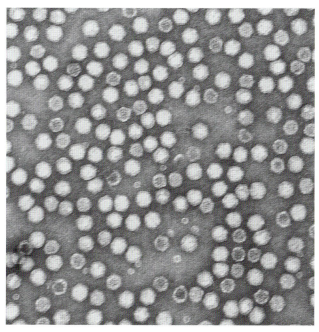

Fig. 140. Particles of *Quail pea mosaic virus*. (Courtesy F. J. Morales)

Fig. 141. Leaf symptoms of bean southern mosaic. (Courtesy F. J. Morales)

Causal Agent

Southern bean mosaic virus (SBMV) is a species of the genus *Sobemovirus* of the family *Tetraviridae*. Virions are approximately 30 nm in diameter and contain a single-stranded RNA molecule 4.1 kb in size, and one capsid protein subunit with a relative molecular weight of 29,000.

Several strains of SBMV that infect beans have been described. The type strain (strain B) infects most common bean cultivars systemically and induces local lesions in others. Strain B can infect a limited number of other legumes but does not infect cowpea. The severe bean mosaic strain, or Mexican strain (strain M), causes more severe symptoms in common bean than does strain B, and it also infects cowpea. An isolate of SBMV from Ghana (strain G), which was isolated originally from cowpea, infects many cowpea cultivars systemically and also induces necrosis in some cultivars of common bean. Resistance-breaking strains from systemically infected bean cultivars resistant to strain B have been described.

SBMV is readily detected by various serological tests, such as the Ouchterlony gel diffusion test. These tests provide a rapid and specific diagnostic test for this virus.

Disease Cycle and Epidemiology

SBMV in leaf extracts is readily sap transmissible, probably because of its stability and high concentration in plant tissue. Roots of plants infected with SBMV release the virus into the soil, and bean seeds planted in this virus-infested soil can acquire the virus and produce virus-infected plants. SBMV can also infect bait plants grown in soil containing virus-infested debris. The virus is seedborne in beans at levels as high as 5%, and this characteristic may be of significance in sporadic outbreaks of the disease. Fresh pods contain a considerable amount of virus both in the pods and seeds. However, most of these virus particles are inactivated during the subsequent drying process as dry bean seeds mature.

Leaf-feeding beetles are efficient vectors of SBMV, and in some cases, the virus is retained by beetles for as long as 2 weeks. Beetle vectors include *Cerotoma trifurcata* (Forster), *Diabrotica adelpha* Harold, *Epilachna varivestis* Mulsant, and other related beetles that feed on beans.

Management

No concerted effort has been made to introduce genetic resistance to SBMV in common bean, because the virus does not usually cause noticeable symptoms. It is also possible that highly susceptible common bean genotypes have been consistently eliminated under field conditions, particularly in breeding programs. Resistance to SBMV in beans is governed by a single gene, and the local lesion reaction is dominant over the

Fig. 142. Rugosing, stunting, and witches'-broom symptoms of bean southern mosaic. (Courtesy F. J. Morales)

systemic type of response. Absolute immunity against the type strain of SBMV in beans is not known, but cultivars that form local lesions are considered resistant for commercial purposes.

Theoretically, the spread of SBMV could be prevented by the use of insecticides or other methods to manage beetles. However, the virus is highly contagious and can be easily spread in common bean fields by workers and contaminated tools.

Selected References

Caldeira, J., and Costa, C. L. 1985. Effects of southern bean mosaic virus infection on bean plant yield. Fitopatol. Bras. 10:310.

Cupertino, F. P., Lin, M. T., Kitajima, E. W., and Costa, C. L. 1982. Occurrence of southern bean mosaic virus in central Brazil. Plant Dis. 66:742-743.

Grogan, R. G., and Kimble, K. A. 1964. The relationship of severe bean mosaic virus from Mexico to southern bean mosaic virus and its related strain in cowpea. Phytopathology 54:75-78.

Teakle, D. S., and Morris, T. J. 1981. Transmission of southern bean mosaic virus from soil to bean seeds. Plant Dis. 65:599-600.

Yerkes, W. D., Jr., and Patiño, G. 1960. The severe bean mosaic virus, a new bean virus from Mexico. Phytopathology 50:334-338.

(Prepared by R. C. Gergerich; Revised by F. J. Morales)

Bean Summer Death

Bean summer death was first described in 1968 as a new disease of common bean in New South Wales, Australia. This disease was very damaging to stringless common bean cultivars grown for processing. The disease is now present in all states of mainland Australia.

Symptoms

Bean summer death is generally manifested as foliar yellowing and plant death, particularly in stringless bean cultivars. Dry bean cultivars are usually less affected.

Causal Agent

Bean summer death is caused by *Tobacco yellow dwarf virus* (TYDV), a species of the genus *Mastrevirus* of the family *Geminiviridae*. Virions are geminate (18 × 35 nm) and contain a circular single-stranded DNA molecule 2.8 kb in size encapsidated from a single protein subunit with a relative molecular weight of 27,500.

Disease Cycle and Epidemiology

TYDV is transmitted by the leafhopper *Orosius argentatus* (Evans) in a persistent manner. The virus is not transmitted mechanically or through seeds, but it can be transmitted by grafting between solanaceous hosts. Natural hosts of TYDV include *Nicotiana tabacum* L., *Beta vulgaris* L., *Datura stramonium* L., *D. tatula* L., and *Lycopersicon esculentum* Mill. Other susceptible hosts are *Medicago sativa* L., *Nicotiana glutinosa* L., *N. rustica* L., *Solanum nigrum* L., *Sonchus oleraceus* L., and *Trifolium repens* L.

Management

The use of resistant common bean genotypes is the best disease management strategy. Great northern UI 31, UI 59, and UI 123; pinto UI 111 and UI 114; 'Red Mexican' UI 34, UI 36, and UI 37; and 'Sanilac' are good sources of resistance to TYDV. Some common bean genotypes possessing resistance to *Beet curly top virus*, another geminivirus transmitted by leafhoppers but belonging to a different genus (*Curtovirus*), also show resistance to TYDV. Management of the leafhopper vector with systemic insecticides at planting time should help prevent early infection of susceptible genotypes.

Selected References

Ballantyne, B. 1970. Field reactions of bean varieties to summer death in 1970. Plant Dis. Rep. 54:903-905.

Thomas, J. E. 1996. Tobacco yellow dwarf monogeminivirus. Pages 1280-1281 in: Viruses of Plants. A. A. Brunt, K. Crabtree, M. J. Dallwitz, A. J. Gibbs, and L. Watson, eds. CAB International, Wallingford, U.K.

Thomas, J. E., and Bowyer, J. W. 1980. Properties of tobacco yellow dwarf and bean summer death viruses. Phytopathology 70:214-217.

(Prepared by F. J. Morales)

Bean Yellow Dwarf

Bean yellow dwarf was described in 1997 as a severe disease of common bean in South Africa. The disease caused yield losses of up to 92% in bean plantings of the cultivar Bonus growing in the Northern Province and Mpumalanga districts.

Symptoms

Systemically infected seedlings exhibit leathery and brittle primary leaves; thick, short internodes with plant stunting; and downward curling of trifoliolate leaves (Fig. 143). Interveinal chlorosis has also been described for other bean genotypes and susceptible plant species using infectious clones under experimental conditions.

Causal Agent

Bean yellow dwarf virus (BeYDV) is a species of the genus *Mastrevirus* of the family *Geminiviridae*, adapted to dicotyledonous hosts. BeYDV consists of a single genomic component of circular single-stranded DNA with 2,561 nucleotides. It is most closely related to *Tobacco yellow dwarf virus* (TYDV), the causal agent of bean summer death in Australia. Sequence identities between these two viruses range from 56.9 to 76.7%. BeYDV is also distantly related to *Chickpea chlorotic dwarf virus*, a tentative mastrevirus that has been shown to infect common bean under experimental conditions in India.

Disease Cycle and Epidemiology

The insect vector of BeYDV is not known, but a leafhopper is suspected. BeYDV was not transmitted by mechanical means to several dicotyledonous species tested.

Management

The selection and use of resistant cultivars is recommended. Preventive measures, such as protection of susceptible common bean cultivars at sowing with a systemic insecticide, might help reduce disease severity.

Selected References

Horn, N. M., Reddy, S. V., Roberts, I. M., and Reddy, D. V. R. 1993. Chickpea chlorotic dwarf virus, a new leafhopper-transmitted geminivirus of chickpea in India. Ann. Appl. Biol. 122:467-469.

Liu, L., Tonder, T. V., Pietersen, G., Davies, J. W., and Stanley, J. 1997. Molecular characterization of a subgroup I geminivirus from a legume in South Africa. J. Gen. Virol. 78:2113-2117.

(Prepared by F. J. Morales)

Bean Yellow Mosaic

A virus associated with bean yellow mosaic was first recovered from the bean cultivar Red Valentine grown in 1931 near Madison, Wisconsin. In the years that followed, bean virus 2, as it was called, was found to infect bean, pea, and a number of leguminous species in several regions of the United States and elsewhere. Currently, bean yellow mosaic occurs worldwide. It has been reported to produce minor crop damage to devastating epidemics, causing considerable losses in yield and quality of the bean crop. Several factors usually determine the severity of this and other viral diseases, e.g., cultivar, plant age, strain of the virus, and environmental conditions.

Symptoms

Typical foliar mosaic consists of contrasting green and yellowish green areas, often accompanied by bright yellow spots, that intensify as the plant ages. Some strains incite only a mild and diffuse chlorotic mottle and limited plant stunting, whereas others cause a coarse mosaic, rugosity, malformation, and severe stunting (Fig. 144). Some cultivars may also exhibit necrotic spots, veinal and apical necroses, wilting, and premature death. Infections initiated after midseason usually only cause mild foliar symptoms. Pods infected while developing exhibit a light green mottle and slight malformation. These are often of diagnostic value in distinguishing *Bean yellow mosaic virus* (BYMV) infection from that caused by *Clover yellow vein virus*. The latter virus, formerly known as the severe strain of BYMV, causes very prominent pod distortion. BYMV-infected plants are usually stunted and bushy because of a reduction of internode length and a proliferation of lateral branches. Their leaves are often cupped downward, similar to plants infected

Fig. 143. Plant stunting and leaf curling caused by *Bean yellow dwarf virus*. (Courtesy F. J. Morales)

Fig. 144. Mosaic and rugosity on leaves caused by *Bean yellow mosaic virus*. (Courtesy F. J. Morales)

73

with *Bean curly top virus*. However, this cupping is more pronounced in older leaves of BYMV-infected plants, whereas the younger leaves of plants with curly top show the most pronounced cupping. Plant maturity is delayed, and the quality and quantity of seed production is frequently affected.

Causal Agent

BYMV is a species of the genus *Potyvirus*. Virus particles are long, flexuous rods 15 × 750 nm and contain a single strand of RNA approximately 10 kb in size. Serologically, BYMV is related to a number of other potyviruses, such as *Bean common mosaic virus*, *Clover yellow vein virus*, *Soybean mosaic virus*, and *Watermelon mosaic virus 2*. BYMV can infect most species of the family Leguminosae, woody plants, and nonlegume species. A number of variants, strains, and pathotypes have been characterized. However, some of the more notable strains are now considered to be distinct viral species.

Disease Cycle and Epidemiology

BYMV is spread from cultivated and wild host reservoirs (*Lathyrus* spp., *Melilotus* spp., and *Trifolium* spp.) by several aphid species in a nonpersistent manner. The virus is acquired after probes of a few minutes and can be retained for several hours. Similarly, infection follows feedings of a few minutes on the host plant, but there is no evidence of a latent period between acquisition and transmission of this virus nor of its multiplication in aphid vectors. More than 20 species of aphids have been reported to transmit BYMV, most notably pea, bean, foxglove, potato, and green peach aphids. Within each of these species are biotypes that may differ in transmission efficiency. The virus is also easily transmitted mechanically. No seed transmission has been demonstrated for BYMV in common bean or any other species of the genus *Phaseolus*. Dissemination of BYMV may also occur through infected vegetatively propagated species, such as *Gladiolus* spp.

Management

Eradication of overwintering hosts of BYMV is usually an impossible task, and applications of insecticides for the management of aphid vectors have not prevented or contained any viral disease. The use of resistant cultivars is the most efficient and economical method of viral disease management. Resistance can significantly reduce losses caused by this virus, particularly in dry bean cultivars. The major source of resistance is derived from an interspecific cross between *Phaseolus coccineus* L. and *P. vulgaris* L. A single dominant gene, *By-2*, confers resistance to most isolates and strains of the virus. Another single dominant gene (*By*) is present in many cultivars of beans. However, this gene confers resistance only to the pea strain of the virus (BYMV-P). A second source of strain-specific resistance was found in the great northern cultivar UI 31. Resistance in this cultivar is conferred by three complementary recessive genes. Resistance conferred by specific genetic factors offers a reliable and durable form of management for many viral diseases. Unquestionably, the incorporation of the *By-2* gene into a large number of bean cultivars could significantly reduce the incidence of bean yellow mosaic.

Selected References

Dickson, M. H., and Natti, J. J. 1968. Inheritance of resistance of *Phaseolus vulgaris* to bean yellow mosaic virus. Phytopathology 58:1450.

Hagel, G. T., Silbernagel, M. J., and Burke, D. W. 1972. Resistance to aphids, mites, and thrips in field beans relative to infection by aphidborne viruses. U.S. Dep. Agric. Bull. Entomol. Res. Div. ARS 33-139.

Hampton, R. O. 1967. Natural spread of viruses infectious to beans. Phytopathology 57:476-481.

Schroeder, W. T., and Provvidenti, R. 1968. Resistance of bean (*Phaseolus vulgaris*) to the PV2 strain of bean yellow mosaic virus conditioned by the single dominant gene *By*. Phytopathology 58:1710.

Swenson, K. G. 1957. Transmission of bean yellow mosaic virus by aphids. J. Econ. Entomol. 50:727-731.

Tatchell, S. P., Baggett, J. R., and Hampton, R. O. 1985. Relationship between resistance to severe and type strains of bean yellow mosaic virus. J. Am. Soc. Hortic. Sci. 110:96-99.

(Prepared by R. Provvidenti; Revised by F. J. Morales)

Bean Yellow Stipple

Bean yellow stipple was first reported in Illinois in 1948 and in Costa Rica in 1972. It occurs sporadically and has caused severe yield losses in the Caribbean and Central America, where it has been associated with the male sterility disease of beans in Costa Rica. The pathogen has also been isolated from naturally infected soybean in Arkansas and cowpea in Costa Rica.

Only leguminous species are known to be systemically infected by the virus. These include several Central and North American cultivars of beans. Species of legumes that exhibit local necrotic lesions and no systemic infection include *Crotalaria juncea* L. and some cultivars of *Glycine max* (L.) Merr. *Dolichos lablab* L. has been routinely used as a local lesion host in infectivity tests of the virus. Infected *Chenopodium album* L. and *C. amaranticolor* Coste & Reyn., nonleguminous indicator species, form white local lesions to the virus.

Symptoms

Leaves of infected bean plants first exhibit a light yellow mottle that gradually develops into clearly distinct small spots that are unevenly distributed in the trifoliolate leaves and form the yellow stippling characteristic of the disease (Fig. 145). The small spots may coalesce, resulting in larger, well-defined, irregularly shaped, bright yellow patches. Both the number and intensity of the spots decrease in younger leaves as plants mature. Severity of symptoms varies according to environmental conditions, time of infection, and bean genotype. Slight stunting of infected plants or necrosis of veins occurs in some cultivars. Reduction of leaf size, rugosity, and deformation have been observed in certain common bean genotypes grown under

Fig. 145. Yellow stippling caused by *Cowpea chlorotic mottle virus*. (Courtesy F. J. Morales)

highland or cool glasshouse conditions. Perhaps the most damaging effect of the virus is its capacity to reduce pod formation and seed production.

Causal Agent

Bean yellow stipple is caused by *Cowpea chlorotic mottle virus* (CCMV), a species of the genus *Bromovirus*. Virions are icosahedral, 26–30 nm in diameter, and contain 23.7% single-stranded RNA in three parts. A subgenomic mRNA (RNA-4) molecule encodes the coat protein. The virus sediments in density gradients as a single centrifugal component at 81 S. CCMV may be serologically related and it is often confused with some cucumoviruses that infect common bean (these genera belong to the same family, *Bromoviridae*).

Disease Cycle and Epidemiology

CCMV is easily sap-transmitted but is not seed-transmitted. Natural spread of the virus occurs via chrysomelid beetles, primarily *Cerotoma ruficornis* (Olivier) and *Diabrotica balteata* LeConte, the two main and most abundant vectors in Central America. They may transmit the virus after acquisition access periods of less than 24 h. The virus is retained for 3–6 days by *C. ruficornis* and for 1–3 days by *D. balteata*. Both species of beetle are inefficient vectors, because only a few individuals may acquire and transmit the virus. However, the large populations of chrysomelids that may be found in bean plantings may cause high virus incidence.

Management

Management of CCMV in bean plantings is aimed at reducing the population densities of its chrysomelid vectors. The frequent association of common bean and maize in Central America favors their development. All common bean cultivars tested to date have shown susceptibility to CCMV.

Selected References

Fulton, J. P., Gámez, R., and Scott, H. A. 1975. Cowpea chlorotic mottle and bean yellow stipple viruses. Phytopathology 65:741-742.
Fulton, J. P., Scott, H. A., and Gámez, R. 1980. Beetles. Pages 115-132 in: Vectors of Plant Pathogens. K. Maramorosch and K. F. Harris, eds. Academic Press, New York.
Gámez, R. 1972. Some properties and beetle transmission of bean yellow stipple virus. (Abstr.) Phytopathology 62:759.
Gámez, R. 1976. Los virus del frijol en Centroamerica, IV: Algunas propiedades y transmision por insectos crisomelidos del virus del moteado amarillo del frijol. Turrialba 26:160-166.
Morales, F. J., and Gámez, R. 1989. Beetle-transmitted viruses. Pages 363-377 in: Bean Production Problems in the Tropics. H. F. Schwartz and M. A. Pastor-Corrales, eds. Centro Internacional de Agricultura Tropical (CIAT), Cali, Colombia.

(Prepared by R. Gámez; Revised by F. J. Morales)

Clover Yellow Vein

Clover yellow vein was first described in white clover grown in Surrey, England, in 1965. However, this disease was described by some in 1948 as a pod-distorting syndrome caused by *Bean yellow mosaic virus* (BYMV) and by others as a necrotic or severe reaction caused by the S strain of BYMV. Clover yellow vein is very destructive to beans and is more widespread in the United States, Japan, and elsewhere than is usually realized.

Symptoms

Symptoms caused by *Clover yellow vein virus* (CYVV) are variable, depending upon the bean cultivar, strain of the virus, time of infection, and environmental conditions. The most prevalent strain causes prominent yellow mosaic, malformation (Figs. 146 and 147), and pronounced plant stunting. A number of cultivars also respond with apical or vein necrosis, premature defoliation, wilting, and plant death. These plants are frequently attacked by soilborne pathogens and develop severe root rot, which can be easily mistaken as the primary disease. While the yellow foliar mosaic resembles that incited by strains of BYMV, pods are severely distorted and mottled (Fig. 148), rendering them worthless for fresh market, canning, or freezing.

Causal Agent

CYVV is currently recognized as a distinct species of the genus *Potyvirus*. It has long, filamentous particles (12 × 760 nm) containing a single strand of RNA approximately 9.5 kb in size. CYVV is related to BYMV molecularly and serologically. It is also related serologically to *Lettuce mosaic virus*, *Soybean mosaic virus*, and *Turnip mosaic virus*. CYVV incites the development of cytoplasmic and nuclear inclusions in cells of infected hosts. The host range of CYVV is broader than that of BYMV, since it can infect most leguminous species.

Fig. 146. Mosaic and leaf malformation caused by *Clover yellow vein virus*. (Courtesy F. J. Morales)

Fig. 147. Prominent yellow mosaic and leaf deformation caused by *Clover yellow vein virus*. (Courtesy R. Provvidenti)

Disease Cycle and Epidemiology

The virus is spread from perennial hosts, especially clover, by aphid vectors in a nonpersistent manner. Several aphid species have been identified as carriers of this virus, notably pea, bean, cotton, foxglove, potato, and green peach aphids. CYVV is also easily transmitted mechanically. Attempts to demonstrate seed transmission of CYVV in common bean, fava bean, pea, and other susceptible species have been unsuccessful.

Management

Clover yellow vein is difficult to manage with insecticides or by eliminating overwintering hosts of the virus. A very effective management method is the use of resistant cultivars, which have been developed from great northern and navy beans. Resistance to CYVV is governed by different genetic factors. Resistance in beans to CYVV was first reported in great northern cultivars. Resistance in the great northern cultivar UI 1140 is governed by a single recessive gene (*by-3*), to which a new symbol (*cyv*) has been assigned. Resistance in the navy bean cultivar Clipper is also monogenically recessive. In great northern UI 31, resistance is conferred by two recessive, complementary genes. Sources of resistance to CYVV are also available in some accessions of *Phaseolus coccineus* L.

Selected References

Pratt, M. J. 1969. Clover yellow vein virus in North America. Plant Dis. Rep. 53:210-212.

Provvidenti, R., and Schroeder, W. T. 1973. Resistance in *Phaseolus vulgaris* to the severe strain of bean yellow mosaic virus. Phytopathology 63:196-197.

Tatchell, S. P., Baggett, J. R., and Hampton, R. O. 1985. Relationship between resistance to severe and type strains of bean yellow mosaic virus. J. Am. Soc. Hortic. Sci. 110:96-99.

Uyeda, I., Takahashi, T., and Shikata, E. 1991. Relatedness of the nucleotide sequence of the 3'-terminal region of clover yellow vein potyvirus RNA to bean yellow mosaic potyvirus RNA. Intervirology 32:234-245.

(Prepared by R. Provvidenti; Revised by F. J. Morales)

Cucumber Mosaic

Cucumber mosaic, first described in 1916, occurs throughout the world and affects some 775 plant species in 365 genera and 85 families. Several strains of the causal virus can induce different symptoms in *Phaseolus vulgaris* L., ranging from mild mosaic to severe plant malformation and total yield loss.

Symptoms

In early stages of growth, plants show systemic symptoms consisting of a prominent leaf epinasty (Fig. 149) followed by a mosaic usually confined to a few leaves; they then apparently recover. Foliar symptoms depend on the cultivar and include leaf curling, green or chlorotic mottle, blisters (Fig. 150), dark green veinbanding, and a zipperlike rugosity along the main veins (Fig. 151). Some cultivars exhibit a leaf deformity that can be confused with that caused by herbicide injury. Although plants may recover from symptoms, the virus continues to replicate in symptomless tissue. In plants that have reached the flowering stage, symptoms, if present, are confined to apical leaves, but pods are mostly curved, mottled, and reduced in size.

Causal Agent

Cucumber mosaic is caused by *Cucumber mosaic virus* (CMV), a species of the genus *Cucumovirus* and consists of 29-nm icosahedral particles with a tripartite genome that is encapsidated in three distinctive particles. Subgenomic mRNA (for coat protein) or satellite RNA may be found in certain CMV strains. Virion protein has one subunit with a relative molecular weight of 24,200. Numerous strains of the virus are recognized. Many of these, including Australian strains CMV-K, CMV-M, CMV-Q, and CMV-U fail to infect common bean by mechanical inoculation. Often, common bean infections attributed to CMV are actually caused by *Peanut stunt virus* (PSV), a different cucumovirus. CMV/PSV hybrids have

Fig. 148. Pod distortion caused by *Clover yellow vein virus.* (Courtesy R. Provvidenti)

Fig. 149. Prominent epinasty caused by *Cucumber mosaic virus.* (Courtesy F. J. Morales)

been detected under natural conditions causing severe plant malformation in common bean.

Disease Cycle and Epidemiology

At least six bean-infecting strains of CMV are seedborne in beans and can thus be disseminated long distances in seed shipments. The Japanese legume strain (CMV-Le) is one of many strains not transmitted in bean seeds. CMV strains infectious to leguminous plants are transmissible in a nonpersistent manner (stylet-borne) by many common species of aphid, the principal natural means of local dissemination. Both seed and aphid transmission of CMV may be erratic and are influenced by factors that are difficult to manipulate experimentally.

CMV can survive from year to year by aphid transmission to perennial, biennial, or overwintering annual weed or crop plant species and by transmission through seeds of several plant species. Survival by any of these means provides CMV inoculum from which aphid vectors can subsequently spread the virus during growth of the crop. In southern France, for instance, CMV occurred naturally in 39 weed and crop species, 13 of which are perennials. Except where prevented by plant-host restrictions on aphid feeding behavior, all alternative hosts of CMV can contain inoculum of the virus that can be transferred to economic hosts. Unusual, massive spread of CMV to a bean planting in New York in 1974 suggested that an abundant source of the virus occurred in that region in conjunction with a large population of aphid vectors.

CMV has not been detected in commercial bean breeding nurseries in the northwestern United States, apparently indicating that neither legume-infecting nor seed-transmitting strains of the virus are present. Most CMV strains known to be seed transmissible in common bean have been detected north and west of the Mediterranean Sea. In this area, large tracts of land are annually planted to common bean, and CMV is transmitted by aphids from many plant species serving as reservoirs of inoculum.

Management

In certain geographic areas (e.g., Japan, southern France, and Turkey), legume crop losses caused by CMV warrant application of specific management measures. Use of virus-free seeds is probably the least costly management measure in a crop production area that is free of bean-infecting strains of CMV. It may also be useful to destroy reservoir hosts or to isolate the crop from them (mainly cucurbits). Management of aphid vector populations in crops or inoculum-reservoir hosts may reduce the natural spread of CMV. Few resistant cultivars are available. However, while no genes are known in common bean to confer immunity to CMV infection, useful levels of tolerance to cucumber mosaic and of resistance to CMV transmission through seeds do exist in this host and are available for disease management through breeding. Several species of the genus *Phaseolus* are resistant to cucumber mosaic, including *P. acutifolius* A. Gray, *P. adenanthus* G. F. W. Mey. (syn. *Vigna adenantha* (G. F. W. Mey.) Marechal, Mascherpa & Stanier), *P. leptostachyus* Benth., *P. polyanthus* Greenman, *P. trilobus* (L.) Ait., and some accessions of *P. coccineus* L. The mechanism of resistance is not known.

Fig. 150. Foliar chlorosis and dark green blisters caused by *Cucumber mosaic virus*. (Courtesy R. Provvidenti)

Fig. 151. Veinbanding, leaflet elongation, and zipperlike rugosity along the main vein caused by *Cucumber mosaic virus*. (Courtesy R. O. Hampton)

Selected References

Bird, J., Sanchez, J., Rodriguez, L., Cortes-Monllor, A., and Kaiser, W. J. 1974. A mosaic of beans (*Phaseolus vulgaris* L.) caused by a strain of common cucumber mosaic virus. J. Agric. Univ. P. R. 53:151-161.

Bos, L., and Maat, D. Z. 1974. A strain of cucumber mosaic virus seed transmitted in bean. Neth. J. Plant Pathol. 80:113-123.

Davis, R. F., and Hampton, R. O. 1986. Cucumber mosaic virus isolates seedborne in *Phaseolus vulgaris*: Serology, host-pathogen relationships, and seed transmission. Phytopathology 76:999-1004.

Horvath, J. 1983. The role of some plants in the ecology of cucumber mosaic virus with special regard to bean. Acta Phytopathol. Acad. Sci. Hung. 18:217-224.

Meiners, J. P., Smith, F. F., and Lawson, R. H. 1974. A strain of cucumber mosaic virus seed-borne in bean (*Phaseolus vulgaris* L.). J. Agric. Univ. P. R. 61:137-147.

Provvidenti, R. 1976. Reaction of *Phaseolus* and *Macroptilium* species to a strain of cucumber mosaic virus. Plant Dis. Rep. 60:289-293.

(Prepared by R. O. Hampton and R. Provvidenti; Revised by F. J. Morales)

Curly Top

Curly top causes mild to severe yellows-type diseases of a large number of hosts in arid and semiarid regions of western North America, the Mediterranean region, Iran, and India. Its occurrence in beans was first reported from California in 1919. A severe epidemic in Twin Falls, Idaho, in 1924 triggered extensive state and federal management programs that persisted into the 1970s. Although a number of resistant cultivars are available, many susceptible cultivars, especially of garden snap beans, are still produced in the western United States seed production areas where sporadic epidemics may cause serious losses.

Symptoms

Symptoms on susceptible cultivars vary greatly since there is a wide range of sensitivity. Symptom expression is also influenced by stage of host growth, prevailing temperatures, and virus strain. The initial symptom on highly sensitive cultivars infected in the crookneck or primary leaf stage is a tight epinastic curl of the first trifoliolate leaves (Fig. 152). The primary leaf also turns yellow, and the plant dies within a few weeks (Fig. 153). Tolerant cultivars or plants infected at successively later stages of growth show progressively less-severe symptoms, and then only on new growth. Stunting is common, and affected leaves are also thicker. Plants infected very late in the season may mature a few seed-bearing pods if those pods were present and fairly well developed before infection. Tolerant or late-infected plants do not always die, but leaves characteristically show an epinastic curl (Fig. 154) and turn brittle and break off readily. Trifoliolate leaves that were already partly formed at infection may remain dark green in a malformed downward curl.

Causal Agent

Beet curly top virus (BCTV) is the type species of the genus *Curtovirus* of the family *Geminiviridae*. BCTV has a circular, single-stranded DNA genome approximately 2.8 kb in size encapsidated in isometric particles 18–22 nm in diameter. Many strains have been reported. It is neither seedborne nor mechanically transmitted and is transmitted only by the leafhoppers

Fig. 152. Epinastic leaf curl and general chlorosis on plant in foreground caused by *Beet curly top virus*. (Courtesy M. J. Silbernagel)

Fig. 153. Distribution of bean plants infected by *Beet curly top virus*. (Courtesy R. L. Forster)

Circulifer tenellus (Baker) in North America and *C. tenellus* and *C. opacipennis* (Lethierry) in other areas.

Disease Cycle and Epidemiology

The disease cycle in western North America is inextricably tied to the life cycle of the leafhopper vector, which is adapted to a hot, dry, desert environment. The vector overwinters as the adult (primarily in desert areas) on a number of BCTV-susceptible hosts, especially the mustard species tumble mustard (*Sisymbrium altissimum* L.) (Fig. 155), flixweed (*Descurainia sophia* (L.) P. Webb ex Prantl), and green tansy mustard (*Descurainia pinnata* (Walt.) Britt.). These winter annuals are usually sprouted by fall rains and serve as a host plant for both the virus and for several cycles of egg laying by the vector during the early spring. By late spring, as desert mustards dry, leafhoppers migrate to nearby irrigated crops. They can, however, migrate great distances, inciting diseases in midwestern and eastern states. Leafhoppers land on a number of green crops, but after initial feeding, they may move from nonpreferred crops, such as bean or tomato, to preferred crops, such as sugar beet, where they remain and multiply for most of the summer.

Fig. 154. Tight epinastic curl of trifoliolate leaves, distorted stem tip, and death of secondary buds caused by *Beet curly top virus*. (Courtesy M. J. Silbernagel)

Fig. 155. Epinastic curl of leaflets, petioles, and inflorescence of tumble mustard (*Sisymbrium altissimum*) caused by *Beet curly top virus*. (Courtesy M. J. Silbernagel)

As these hosts mature or are harvested in the fall, adult leafhoppers may be found in large numbers on Russian thistle (*Salsola kali* L. var. *tenuifolia* Tausch) while it is still green and succulent. Leafhoppers return to desert overwintering areas or to nearby uncropped areas to feed on rain-sprouted wild mustards.

Management

Growing resistant or tolerant cultivars is the most effective means of management, followed by avoidance of proximity to overwintering areas. In the 1930s, massive spray programs were initiated in the foothills of California and desert areas of southern Idaho to kill the vectors. The host mustard species were also replaced with nonhost grass species. Those programs were fairly successful but were discontinued by the early 1970s. Late planting of susceptible cultivars (after the major leafhopper migrations are finished) can also afford some reduction in damage. Use of insecticides to manage the vector on susceptible bean crops is largely ineffective in managing the disease.

Resistance in beans is attributable to two major epistatic genes, one dominant and one recessive. Unpublished studies by Silbernagel indicate two dominant major genes for host resistance to the virus and a number of minor genes that may influence vector preference.

Selected References

Bennett, C. W. 1971. The Curly Top Disease of Sugarbeet and Other Plants. Monogr. 7. American Phytopathological Society, St. Paul, MN.

Larsen, R. C., and Miklas, P. N. 2004. Generation and molecular mapping of a sequence characterized amplified region marker linked with the *Bct* gene for resistance to *Beet curly top virus* in common bean. Phytopathology 94:320-325.

Silbernagel, M. J. 1965. Differential tolerance to curly top in some snap bean varieties. Plant Dis. Rep. 49:475-477.

Silbernagel, M. J., and Jafri, A. M. 1974. Temperature effects on curly top resistance in *Phaseolus vulgaris*. Phytopathology 64:825-827.

Thomas, P. E. 1972. Mode of expression of host preference by *Circulifer tenellus*, the vector of curly top virus. J. Econ. Entomol. 65:119-123.

Thomas, P. E., and Mink, G. I. 1979. Beet curly top virus. Descriptions of Plant Viruses, No. 210. Commonwealth Mycological Institute and Association of Applied Biologists, Kew, Surrey, England.

(Prepared by M. J. Silbernagel; Revised by F. J. Morales)

Peanut Mottle

Peanut mottle is a severe disease of navy beans first reported from Queensland, Australia, in 1972. In 1983 and 1985, a similar disease was observed in central New York and in west Texas, respectively. In Texas, lethal systemic necrosis of processing snap beans caused significant economic losses between 1985 and 1987.

Symptoms

Three major symptoms occur in beans. Necrotic lesions and vein necrosis may occur on mechanically inoculated cotyledonary leaves of susceptible plants (Fig. 156); uninoculated trifoliolate leaves develop mosaic, which is followed quickly by chlorosis and necrosis of veins, petioles, and stems (Fig. 157). These plants frequently die. Similar symptoms may occur under field conditions and are sometimes mistaken for the systemic necrosis (black root) induced by *Bean common mosaic virus* and *Bean common mosaic necrosis virus*. Other genotypes may react with necrotic lesions and vein necrosis (Fig. 156) or with chlorotic lesions that eventually coalesce (Fig. 158). With both reactions, the virus remains localized in inoculated leaves. Virus concentrations in localized chlorotic infections in bean leaves and in susceptible infected leaves of peanut and soybean can be similar. The necrotic symptoms are usually the result of a hypersensitive reaction of common bean genotypes possessing monogenic dominant resistance to *Bean common mosaic virus* and related viruses, such as the causal virus of peanut mottle.

Causal Agent

The causal agent is *Peanut mottle virus* (PeMoV), a species of the genus *Potyvirus*. In addition to beans, PeMoV occurs in nature in several important legume crops. Particles are flexuous

Fig. 156. Necrotic lesions and vein necrosis on a primary leaf caused by *Peanut mottle virus*. (Courtesy C. W. Kuhn)

Fig. 157. Chlorosis and necrosis of veins and petioles of a trifoliolate leaf caused by *Peanut mottle virus*. (Courtesy C. W. Kuhn)

Fig. 158. Chlorotic lesions on a leaf of a resistant plant inoculated with *Peanut mottle virus*. (Courtesy C. W. Kuhn)

rods, 12 × 750 nm. They contain one molecule of positive-sense, single-stranded RNA that has a molecular weight of 3.0×10^6. The RNA is encapsidated with a single type of capsid protein subunit with a relative molecular weight of 35,000. Significant quantities (50–150 μg per gram of infected tissue) of highly purified virus can be obtained from diseased pea plants. PeMoV is moderately immunogenic and can be evaluated with standard serological tests. Immunosorbent electron microscopy and enzyme-linked immunosorbent assays have demonstrated a serological relationship to a few other potyviruses.

Disease Cycle and Epidemiology

PeMoV is easily transmitted mechanically to and from bean plants when infected tissue is ground in buffer (pH 7–8). For some infected species, particularly peanut, it is desirable to add antioxidants, such as sodium bisulfite and sodium diethyldithiocarbamate, to the buffer.

Seed transmission of PeMoV occurs in low frequencies in bean, cowpea, lupin, and peanut. No seed transmission in soybean or infected weed hosts has been reported in nature. Several aphid species can transmit PeMoV readily in a nonpersistent manner. Vector transmission studies have not been conducted with beans, but presumably bean infections in nature have resulted from aphid transmission.

Peanut plants grown from seeds infected with PeMoV appear to provide the source of primary inoculum for infection of susceptible crops. This source has been established for cowpea, peanut, and soybean, and the occurrence of PeMoV in beans in Australia and the United States (New York and Texas) was associated with peanut plants grown in nearby fields. In the United States, approximately 200,000 peanut seeds are planted per hectare. Therefore, a seed transmission frequency as low as 0.1% would provide about two infected seedlings per 100 m² in a field. When aphids are present, PeMoV can spread within a field and to nearby susceptible crops.

Management

PeMoV is an opportunistic pathogen of common bean and, consequently, peanut mottle should be managed successfully by isolation of bean crops from peanut crops. Studies in the United States have shown that very little or no infection develops in susceptible plants that are grown 100 m or more away from infected peanut plants.

More than 65 bean cultivars evaluated in New York were resistant to PeMoV (local infection only, no systemic infection). Resistance associated with necrotic local lesions is conditioned by a single but incompletely dominant gene. Other studies have demonstrated that none of eight variants of PeMoV could overcome either the necrosis or chlorosis localization types of resistant reactions in beans. Many horticulturally acceptable resistant bean cultivars are currently available, and a breeding program could be used to develop new cultivars.

Selected References

Behncken, G. M., and McCarthy, G. J. P. 1973. Peanut mottle virus in peanuts, navy beans and soybeans. Qld. Agric. J. 99:636-639.
Bock, K. R., Guthrie, E. J., and Meredith, G. 1978. Viruses occurring in East Africa that are related to peanut mottle virus. Ann. Appl. Biol. 89:423-428.
Kuhn, C. W. 1965. Symptomatology, host range, and effect on yield of a seed-transmitted peanut virus. Phytopathology 55:880-884.
Provvidenti, R., and Chirco, E. M. 1987. Inheritance of resistance to peanut mottle virus in *Phaseolus vulgaris*. J. Hered. 78:402-403.
Thomas, J. E., Redden, R. J., and Usher, T. 1990. Screening for resistance to *Peanut mottle virus* in accessions and breeding lines of *Phaseolus vulgaris*. Bean Improv. Coop. Rep. 33:163-164.

(Prepared by C. W. Kuhn; Revised by F. J. Morales)

Peanut Stunt

Peanut stunt was first isolated from peanut (*Arachis hypogaea* L.) in Virginia in the mid-1960s. It affects several legume crops, including common bean, alfalfa, and several species of clover, and has been reported to be economically damaging on beans in Africa (Morocco and Sudan), Europe (France and Spain), Turkey, Japan, the United States (North Carolina and Virginia), and Chile.

Symptoms

Symptoms on field-grown beans may include mosaic, systemic necrosis, leaf rugosity, deformation (Fig. 159), and stunting (Fig. 160). Pods are scarce, small, and malformed; and few seeds are produced. Greenhouse-grown plants may show chlorotic or necrotic local lesions and usually show epinasty and varying degrees of mosaic and malformation depending on cultivar and strain. Sometimes the trifoliolate leaves are elongated and pointed (Fig. 161), resembling herbicide (2,4-D) injury. In many cases, symptoms could be confused with those produced by severe strains of *Bean yellow mosaic virus*.

Causal Agent

Peanut stunt virus (PSV) is a species of the genus *Cucumovirus* of the family *Bromoviridae*. Its genome is composed of three positive-sense, single-stranded RNAs (3.4, 2.9, and 2.2 kb in size) encapsulated in isometric particles about 30 nm in diameter. Subgenomic mRNA (RNA-4), which encodes the coat protein, and satellite RNA are nongenomic viral nucleic

Fig. 159. Mosaic (left), rugosity, and deformation (right) of cultivar Eagle leaves caused by *Peanut stunt virus*. (Courtesy J. R. Stavely)

Fig. 160. Stunting of a cultivar BBL 283 bean plant caused by *Peanut stunt virus* (right). Healthy plant is on the left. (Courtesy J. R. Stavely, from the files of J. P. Meiners)

acids. PSV is readily transmitted mechanically and by several aphid species in a nonpersistent manner. In addition to a wide leguminous host range, one or more species of the families Chenopodiaceae, Compositae, Cucurbitaceae, and Solanaceae are also susceptible. Many strains have been reported, which vary in host range, serological relationships, and particle stability. A natural hybrid recombining RNA 1 and RNA 2 from PSV and RNA 3 from *Cucumber mosaic virus* was found affecting common bean in Chile.

Disease Cycle and Epidemiology

The epidemiology of peanut stunt is characterized by sudden disease outbreaks of unexplained origin. Although the virus is reportedly seed transmissible at very low rates, infected plants rarely produce pods with normal viable seeds. Thus, bean seeds are unlikely to be a source of primary inoculum. Susceptible and widespread perennial hosts, such as alfalfa and clover, are a more likely source of inoculum. Since natural spread depends upon aphid vectors, conditions that cause aphid populations to increase in lucerne and clover or to migrate to beans may be key factors in epidemics of peanut stunt in bean crops.

Management

There are no known sources of resistance in common bean, and no specific management measures have been developed. However, avoidance of proximity to infected leguminous forage crops (alfalfa and clover) or seed crops (peanut and soybean) may be helpful. It is doubtful if spraying the bean crop with insecticides would greatly reduce disease incidence in beans.

Selected References

Diaz-Ruiz, J. R., Kaper, J. M., Waterworth, H. E., and Devergne, J. C. 1979. Isolation and characterization of peanut stunt virus from alfalfa in Spain. Phytopathology 69:504-509.

Echandi, E., and Hebert, T. T. 1971. Stunt of beans incited by peanut stunt virus. Phytopathology 61:328-330.

Lot, H., and Kaper, J. M. 1976. Physical and chemical differentiation of three strains of cucumber mosaic virus and peanut stunt virus. Virology 74:207-222.

Milbrath, G. M., and Tolin, S. A. 1977. Identification, host range and serology of peanut stunt virus isolated from soybean. Plant Dis. Rep. 61:637-640.

Mink, G. I., Silbernagel, M. J., and Saksena, K. N. 1969. Host range, purification, and properties of the western strain of peanut stunt virus. Phytopathology 59:1625-1631.

Waterworth, H. E., Monroe, R. L., and Kahn, R. P. 1973. Improved purification procedure for peanut stunt virus, incitant of Tephrosia yellow vein disease. Phytopathology 63:93-98.

White, P. S., Morales, F. J., and Roossinck, M. J. 1995. Interspecific reassortment of genomic segments in the evolution of cucumoviruses. Virology 207:334-337.

(Prepared by M. J. Silbernagel; Revised by F. J. Morales)

Red Node

Red node was initially observed in beans in 1938 in northeastern Colorado and was given its current name in 1943. Red node has been reported from the bean-growing areas of the western United States, where it is a disease of minor economic importance. However, occasionally the disease has caused serious losses in commercial bean plantings in some areas.

Symptoms

The initial symptom of red node in beans is usually a reddening of nodes of the stem (Fig. 162) and pulvini of leaves and leaflets. At times, these symptoms are accompanied by necrosis and reddening of veins of primary and trifoliolate leaves (Fig. 163).

Fig. 162. Reddening of nodes caused by the bean red node strain of *Tobacco streak virus*. (Courtesy R. L. Forster)

Fig. 161. Leaf elongation and veinclearing caused by *Peanut stunt virus* (right). Healthy plant is on the left. (Courtesy M. J. Silbernagel)

Fig. 163. Reddening and necrosis of veins of trifoliolate leaves caused by the bean red node strain of *Tobacco streak virus*. (Courtesy H. F. Schwartz)

Discoloration and necrosis of the stem and apical tissues may follow. Infected plants may bend and break at discolored nodes. Pods may develop individual red ring spots several millimeters in diameter or masses of sunken, reddish, pock-mark-like lesions. When necrosis is extensive, pods may become shriveled and deformed. Seeds from symptomatic pods are often shriveled and discolored. Plants affected by red node are frequently stunted, and mortality can be high, particularly if plants are infected from seeds or in the seedling and prebloom stages of growth.

Causal Agent

Tobacco streak virus (TSV) is the type species of the genus *Ilarvirus* of the family *Bromoviridae* and consists of isometric RNA-containing particles, 27, 30, and 35 nm in diameter. The virus has a tripartite genome of single-stranded RNA, with one part being present in each of the three main particle types. To be infective, the three-part genome (RNA-1, RNA-2, and RNA-3) must be present, together with either a fourth subgenomic RNA (RNA-4) or coat protein. Infectivity of the TSV genome can be activated with the coat protein of *Alfalfa mosaic virus*, and vice versa.

Several strains of TSV differing in host range, symptomatology, physical properties, or serology have been isolated from different naturally infected plant species worldwide. The bean red node strain is recognized as the causal agent of red node of common bean.

Strains of TSV are reported to be seedborne in beans, as well as in several experimentally and naturally infected plant species. The red node strain was seedborne in an average of 1.4% of 21,424 seeds of several pinto cultivars. The common bean cultivar Black Turtle Soup was infected naturally by two serologically distinct strains of TSV in Washington, one of which produced typical red node symptoms and the other produced mosaic symptoms in beans. Seed transmission of TSV was detected in more than 2% of the 'Black Turtle Soup' bean plants that were naturally infected with the red node strain, but no seed transmission was observed in plants infected with the mosaic-inducing strain. When seed transmission of the red node strain occurred in 'Black Turtle Soup' bean plants, usually more than 90% of the seeds were infected with the virus. Most plants infected from seeds with TSV were stunted and exhibited typical red node symptoms.

In 1975, thrips (*Frankliniella* spp.) were identified as a vector of a Brazilian strain of TSV. Subsequently, the red node strain was transmitted from naturally infected white sweetclover (*Melilotus alba* Medik.) to *Chenopodium quinoa* Willd. and *M. alba* with a mixture of thrips (*Frankliniella occidentalis* (Pergande) and *Thrips tabaci* (Lindeman)) but not by pea aphids or pea leaf weevils.

Disease Cycle and Epidemiology

Although several annual and perennial weed species have been identified as natural hosts of different TSV strains, information is lacking on the reservoir hosts of the bean red node strain. More than 30 years ago, it was observed that the incidence of red node was higher in areas of bean fields that were near white sweetclover (*M. alba*) and yellow sweetclover (*M. officinalis* (L.) Lam.). In 1982, *M. alba* was identified as the primary reservoir of two strains of TSV in eastern Washington, one of which induced red node symptoms in 'Bountiful' and 'Black Turtle Soup' beans. The red node strain was transmitted in 0.9–24% of the seeds of naturally infected *M. alba*. Seed transmission of the virus in *M. alba* may play an important role in the epidemiology of red node in some areas.

Management

Because it is a relatively minor disease of beans, little research has been done on the management of red node. If red node should pose a threat to bean production, certain precau-

tions could be taken to reduce disease losses; seed agencies could then adopt inspection and certification standards for red node. Destruction of potential reservoirs of red node (such as white sweetclover) in close proximity to bean fields may reduce spread of the virus into beans by aerial vectors. Bean cultivars resistant to red node are not generally available.

Selected References

Fulton, R. W. 1983. Ilarvirus group. Descriptions of Plant Viruses, No. 275. Commonwealth Mycological Institute and Association of Applied Biologists, Kew, Surrey, England.

Fulton, R. W. 1985. Tobacco streak virus. Descriptions of Plant Viruses, No. 307. Commonwealth Mycological Institute and Association of Applied Biologists, Kew, Surrey, England.

Kaiser, W. J., Wyatt, S. D., and Pesho, G. R. 1982. Natural hosts and vectors of tobacco streak virus in eastern Washington. Phytopathology 72:1508-1512.

Mink, G. I., Saksena, K. N., and Silbernagel, M. J. 1966. Purification of the bean red node strain of tobacco streak virus. Phytopathology 56:645-649.

Thomas, W. D., Jr., and Graham, R. W. 1951. Seed transmission of red node virus in pinto beans. Phytopathology 41:959-962.

(Prepared by W. J. Kaiser; Revised by F. J. Morales)

Soybean Mosaic

Soybean mosaic sporadically affects common bean under natural conditions, particularly in countries such as Brazil where soybean and common bean are planted extensively in the same regions. The susceptibility of common bean to the causal agent of soybean mosaic had been documented as early as 1948, and the virus is still occasionally isolated from diseased common bean plants in the field.

Symptoms

Soybean mosaic induces three types of symptoms in common bean: mosaic, necrotic and chlorotic local lesions, and systemic necrosis. The mild mosaic and leaf curling symptoms induced by the causal virus of soybean mosaic in common bean are similar to the reactions induced by *Bean common mosaic virus* (BCMV) in some common bean genotypes. However, common bean leaves systemically infected by the causal virus of soybean mosaic usually turn chlorotic at an early stage (Fig. 164). This symptom is known as mosaico-em-colherinha (teaspoon mosaic) in Brazil, because of the distinctive concave foliar curling observed. Local necrotic and chlorotic reactions are also observed but mainly on artificially inoculated plants. Systemic or top necrosis seems to be a hypersensitive reaction of common bean genotypes possessing monogenic dominant resistance to BCMV when infected by the causal virus of soybean mosaic.

Causal Agent

Soybean mosaic is caused by *Soybean mosaic virus* (SMV), a species of the genus *Potyvirus* of the family *Potyviridae*. Virions are typically flexuous filaments approximately 700 nm long and 16 nm wide and contain one single-stranded RNA species 10.4 kb in size. The virus induces the formation of cylindrical inclusions of various types in the cytoplasm of infected plant cells.

Disease Cycle and Epidemiology

Soybean mosaic is not a serious or ubiquitous disease of common bean, although aphids can move the virus effectively between soybean and common bean plantings grown in the same area. SMV can also be transmitted via the seeds of systemically infected common bean plants and is readily mechanically transmissible.

Management

There are no specific recommendations regarding the management of SMV in common bean plantings other than isolating common bean plantings from SMV-susceptible soybean fields. A single dominant gene, *Hss*, that conditions a hypersensitive systemic necrosis response to SMV has been identified in *Phaseolus vulgaris* L.

Selected References

Castaño, M., and Morales, F. J. 1983. Seed transmission of soybean mosaic virus in *Phaseolus vulgaris* L. Fitopatol. Bras. 8:103-107.

Costa, A. S. 1987. Fitoviroses do feijoeiro no Brasil. Pages 173-256 in: Feijão: Fatores de Produção e Qualidade. E. A. Bulisani, ed. Fundacão Cargill, Campinas, São Paulo, Brazil.

Goodman, R., Edwardson, J. R., and Bos, L. 1996. Soybean mosaic potyvirus. Pages 1162-1165 in: Viruses of Plants. A. A. Brunt, K. Crabtree, M. J. Dallwitz, A. J. Gibbs, and L. Watson, eds. CAB International, Wallingford, U.K.

Kyle, M. M., and Provvidenti, R. 1993. Inheritance of resistance to potyviruses in *Phaseolus vulgaris* L. II. Linkage relations and utility of a dominant gene for lethal systemic necrosis to soybean mosaic virus. Theor. Appl. Genet. 86:189-196.

(Prepared by F. J. Morales)

Stipple Streak

Stipple streak of common bean was first described as a virus disease in the Netherlands in 1935. Since then, it has been reported in Germany, Australia (Queensland), and the United States (New York). It has also been detected in glasshouse-grown beans in New York and New Zealand and may cause economic losses under both field and glasshouse conditions. The malady is likely to be more prevalent than realized, since its symptoms may easily be attributed to other causes. Stipple streak may cause economic damage in crops grown in the field and glasshouse.

Symptoms

The disease derives its name from the scattered reddish, dark brown, or black spots and the streaks or bands that occur on stems, petioles, and veins. In addition, reticulate necrosis of veins may develop on leaves in isolated patches differing in size and shape (Fig. 165). Such leaves are malformed and often wither and drop. Pods of infected plants often show rusty brown, irregularly shaped, sometimes concentric necrotic lesions; if affected when young, they may become distorted and shrivel (Fig. 166). Necrotic lesions may develop on pods after harvest.

Distribution of symptoms in plants is often one sided or erratic. Tip leaves sometimes remain normal, and normal pods may be produced. Reports differ as to whether the virus can be isolated from symptomless plant parts. Plant growth is usually curtailed and plants may die prematurely.

Mechanical inoculation of primary leaves leads to many chlorotic, dry, or necrotic local lesions 1–3 days after inoculation. In certain bean genotypes, such lesions may coalesce, leading to desiccation (and sometimes abscission) of leaves. In other genotypes, mechanical inoculation leads to gradually

Fig. 165. Reticulate veinal necrosis symptoms of stipple streak, caused by *Tobacco necrosis virus*. (Courtesy L. Bos)

Fig. 164. Leaf symptoms of soybean mosaic. (Courtesy F. J. Morales)

Fig. 166. Pod necrosis symptoms of stipple streak, caused by *Tobacco necrosis virus*. (Courtesy L. Bos)

extending veinal necrosis in inoculated leaves and systemic disease as described above.

The disease is more common in plants growing in wet soils, particularly peat soils. The symptoms are most severe during warm, cloudy weather. Symptoms may not express in leaves developing during hot, sunny periods. Symptoms may reappear in pods formed later on such plants.

Leaf and stem symptoms may resemble those of bacterial infections (necrosis) or of black root disease caused by *Bean common mosaic virus*. With *Bean common mosaic virus*, leaf and stem systemic necroses are more uniform, and necrotic lesions on affected pods are more internal (including vascular necrosis) than with stipple streak.

Causal Agent

Stipple streak is caused by strains of *Tobacco necrosis virus* (TNV), a species of the genus *Necrovirus* of the family *Tombusviridae*. TNV is an RNA-containing virus approximately 3.8 kb in size, with a protein subunit that has a relative molecular weight of 29,000–30,000. The virus has isometric particles of about 28 nm in diameter. Several strains, including the S strain, may have a 17-nm spherical satellite virus (SV) that is serologically distinct, codes for its own protein, and depends on TNV for its multiplication. When freed from its SV, TNV-S caused milder symptoms in beans and caused systemic mottle with little necrosis in 'Prince' bean plants in the glasshouse. The virus is physically and chemically very stable. It withstands heating for 10 min up to 85–95°C, dilution up to 10^{-4} to 10^{-6}, and storage at room temperature for 2–3 months.

TNV has a wide host range, including several wild species, and its host plants are usually not systemically infected. The virus is readily recognized by its rapid and exclusively local reaction to sap inoculation on cucumber cotyledons and foliage leaves of *Chenopodium amaranticolor* Coste & Reyn., *C. quinoa* Willd., and *Nicotiana tabacum* L. and by its local and systemic reactions on plants of *Phaseolus* beans. Rapid confirmation is by serology (using antisera to strain S, as well as A, B, and C of TNV serogroup A). Other serologically distinct strains of TNV cause ABC disease in potato, augusta disease in tulip, and one form of cucumber necrosis.

Disease Cycle and Epidemiology

The virus is transmitted in soil in a nonpersistent manner by attachment to the surface of zoospores of the root-infecting chytridiomycete fungus *Olpidium brassicae* (Woronin) P. A.

Dang. Thus, dissemination of the virus can occur in irrigation water or splashing rain. Successful attachment of the virus to the fungus depends on the right combination of virus strain and fungus race. Transmission of the virus may occur through plant-to-plant contact. Pods may be infected during harvest and develop symptoms within a few days. Seed transmission is not known.

The virus multiplies rapidly at low light intensities, which may explain why symptoms appear more readily in cloudy than in sunny weather, in shaded than in exposed leaves, and in pole or string beans than in bush beans.

Management

Scarlet runner bean (*Phaseolus coccineus* L.) is practically immune. All common bean cultivars are susceptible and are likely to suffer when grown on infested soil and exposed to weather conditions favorable to the disease. So far, problems with screening by mechanical inoculation have hindered efforts to breed for resistance. Selection under field conditions may give variable results because of unpredictable behavior of the vectors.

Crop rotation to reduce inoculum pressure is of little value because of the wide natural host ranges of both the virus and the vector. Transmission no longer occurs when infested soil is sterilized by heating at 100°C for 30 min, and the same may hold for chemical treatment of soil as for other plant viruses transmitted by fungi. However, chemical management may be too costly for field-grown crops.

Selected References

Behncken, G. M. 1986. Stipple streak disease of French bean caused by a tobacco necrosis virus in Queensland. Aust. J. Agric. Res. 19:731-738.

Horvath, J., and Beczner, L. 1982. Virus susceptibility and virus resistance in bean (*Phaseolus* L.) species: New hosts and new sources of resistance. Kertgazdaslig 14:39-45.

Kassanis, B. 1970. Tobacco necrosis virus. Descriptions of Plant Viruses, No. 14. Commonwealth Mycological Institute and Association of Applied Biologists, Kew, Surrey, England.

Natti, J. J. 1959. A systemic disease of beans caused by a tobacco necrosis virus. Plant Dis. Rep. 43:640-644.

Thomas, W. 1973. A necrotic disease of glasshouse beans caused by bean stipple streak virus. N. Z. J. Agric. Res. 16:150-154.

(Prepared by L. Bos; Revised by F. J. Morales)

Diseases Caused by Phytoplasmas

Phytoplasmas, previously known as mycoplasmalike organisms (MLOs), infect beans and cause yellows diseases (chlorosis), stunting, witches'-broom (excessive proliferation of branches), stem elongation, proliferation of buds, and disorders of reproductive organs, such as virescence (greening) and phyllody (becoming leaflike). Under field conditions, phytoplasmas of beans have been associated with witches'-broom in Japan, Brazil, and El Salvador; machismo disease in Colombia; and phyllody in the United States (Washington).

Phytoplasmas are prokaryotes, lack a cell wall but possess a delimiting cell membrane, are highly variable in shape (pleomorphic), are 0.2–1.0 μm in diameter, and contain ribosomes, RNA, and DNA. They occur in plant sieve elements and phloem parenchyma and are transmitted by leafhoppers. Phytoplasmas are resistant to penicillin but sensitive to tetracycline.

Long Stem

A new disease of common bean characterized by branch proliferation and excessive stem elongation was recently observed in El Salvador. The incidence of the disease was very low, but the unusual plant malformation warranted the characterization of the causal agent.

Symptoms

Diseased common bean plants show characteristic witches'-broom symptoms, accompanied by virescence and phyllody. The most striking symptom, however, is the unusual elongation of the stems to approximately five times the normal length.

Causal Agent

The causal agent was shown by polymerase chain reaction amplification of 16S rDNA to be a phytoplasma of the aster yellows group. The identity of the vector has not yet been determined.

Disease Cycle and Epidemiology

Diseased common bean plants were found in pairs within the affected field planted in relay with maize and next to a papaya field. The possible role of these two plant species in the emergence of this disease in common bean is not known.

Management

Because of the low incidence and recent emergence of this disease in common bean, no specific management measures have been adopted.

(Prepared by R. Hall and F. J. Morales)

Machismo

Machismo was the name given to a disease of soybean observed in the Cauca Valley of Colombia in 1968. In 1981, similar symptoms were observed on common bean plants in the same region, reaching incidences of up to 15%. Its incidence during 1981–1985 was less than 1%. The disease in common bean was also referred to as machismo because flowers are converted into vegetative structures by the causal agent.

Symptoms

Symptoms appear during reproductive development of the plant. Following early infection, flower petals turn light to dark green (virescence) and sepals are elongated. A corrugated structure resembling a rolled leaf (Fig. 167) emerges from the unopened floral apex (phyllody). Pods may become thin, stiff, upright, twisted, corrugated, and crescent shaped (Fig. 168) and set few seeds. Large petioles bearing small flower buds and small leaves on small petioles may develop. The plant adopts the witches'-broom appearance (Fig. 169). Late infection may promote premature viviparous germination of seeds in healthy-appearing pods (Fig. 170).

Causal Agent

The presence of phytoplasmas in the phloem of affected organs has been demonstrated by electron microscopy.

Disease Cycle and Epidemiology

The phytoplasma is transmitted by the brown leafhopper *Scaphytopius fuliginosus* (Osborn) and can infect common bean, lima bean, soybean, *Cajanus cajan* (L.) Millsp., *Crotalaria juncea* L., *Crotalaria spectabilis* Roth, a *Desmodium* sp., *Galactia glaucescens* Kunth., *Rhynchosia minima* (L.) DC., *Vigna angularis* (Willd.) Ohwi & H. Ohashi, *Vigna umbellata* (Thunb.) Ohwi & H. Ohashi, and *Vinca rosea* L. Symptoms appeared 31–43 days after exposure of 1- to 6-day-old bean seedlings to infective adults of *S. fuliginosus* for 5 days. The phytoplasma can be transmitted by grafting but not mechanically or by seeds.

Fig. 167. Phyllody caused by infection with a phytoplasma. (Courtesy H. F. Schwartz)

Fig. 169. Witches'-broom caused by a phytoplasma. (Courtesy H. F. Schwartz)

Fig. 168. Pod deformation caused by a phytoplasma. (Courtesy H. F. Schwartz)

Fig. 170. Viviparous germination of seeds in a pod caused by a phytoplasma. (Courtesy H. F. Schwartz, from the files of G. A. Granada)

Management

Management of machismo is based on planting at recommended dates, maintaining adequate crop rotations, and not cropping continuously or simultaneously with susceptible plants. Infected crop plants and nearby weed hosts should be removed and destroyed. Common bean germ plasm has not been evaluated in Colombia for resistance to machismo. Spraying of insecticides to manage the leafhopper or of oxytetracycline to suppress the pathogen may also reduce the disease incidence but are not currently recommended.

Selected References

Fletcher, J., Irwin, M. E., Bradfute, O. E., and Granada, G. A. 1984. Discovery of a mycoplasmalike organism associated with diseased soybeans in Mexico. Plant Dis. 68:994-996.

Granada, G. A. 1978. Machismo en frijol. ASCOLFI Inf. (Colombia) 4(1):2.

Granada, G. A. 1979. Machismo, a new disease of beans in Colombia. (Abstr.) Phytopathology 69:1029.

Granada, G., and Kitajima, E. 1989. Mycoplasma-like diseases. Pages 321-332 in: Bean Production Problems in the Tropics, 2nd ed. H. F. Schwartz and M. A. Pastor-Corrales, eds. Centro Internacional de Agricultura Tropical (CIAT), Cali, Colombia.

(Prepared by R. Hall and F. J. Morales)

Phyllody

This new disease of bean was recently detected in the Columbia Basin of Washington in dry bean cultivars of Andean origin. Symptoms of dry bean phyllody were observed during mid to late pod development as leafy petals (phyllody) and aborted seed pods resembling thin, twisted, and corrugated leaflike structures. Deformed sterile pods were small, sickle shaped, upright, and leathery. Infected plants generally exhibited chlorosis, stunting, or bud proliferation from leaf axils.

Phytoplasma infection was confirmed with polymerase chain reaction and restriction fragment length polymorphism analyses, and the analyses confirmed that the phytoplasma belonged to the clover proliferation group (16SrVI) subgroup A (16SrVI-A).

This subgroup currently consists of three members, clover proliferation, potato witches'-broom, and vinca virescence.

The relationship of dry bean phyllody to other phytoplasma diseases of beans (i.e., long stem, machismo, and witches'-broom) is unknown, and no management recommendations are reported.

Selected Reference

Lee, I.-M., Bottner, K. D., Miklas, P. N., and Pastor-Corrales, M. A. 2004. Clover proliferation group (16SrVI) subgroup A (16SrVI-A) phytoplasma is a probable causal agent of dry bean phyllody disease in Washington. Plant Dis. 88:429.

(Prepared by H. F. Schwartz)

Witches'-Broom

A witches'-broom disease occurs on beans in Japan. Diseased plants show chlorosis, reduced leaflet size, shoot proliferation, and phyllody of floral organs. Symptoms appear 1 month after feeding by the leafhopper vector *Orosius orientalis* (Matsumura). Little is known concerning its biology or management. Witches'-broom symptoms on common bean are referred to as superbrotamento (branch proliferation) in Brazil. The causal phytoplasma has been visualized by electron microscopy, but neither its identity nor its vector are known. No specific disease management methods have been recommended in Brazil, other than to separate other legumes, such as soybean and *Crotalaria* spp., from common bean plantings.

Selected References

Costa, A. S. 1987. Fitoviroses do feijoeiro no Brasil. Pages 173-256 in: Feijão: Fatores de Produção e Qualidade. E. A. Bulisani, ed. Fundacão Cargill, Campinas, São Paulo, Brazil.

Murayama, D. 1966. On the witches' broom disease of sweet potato and leguminous plants in the Ryukyu Islands. Mem. Fac. Agric. Hokkaido Univ. 6:81-103.

(Prepared by R. Hall and F. J. Morales)

Part II. Noninfectious Diseases

Environmental and Genetic Disorders

Air Pollution

Beans may be damaged by airborne toxic materials originating from urban areas, vehicles, certain industries, and natural environmental processes. Air pollutants that affect beans include ozone, peroxyacetyl nitrate (PAN), sulfur dioxide, fluorides, solid particles, and chlorine. Air pollutants can also influence the interactions between beans and plant pathogens.

Ozone is a common air pollutant that can be formed by electrical discharge during thunderstorms. However, by far the most important source of phytotoxic ozone is the photochemical production from gases liberated by combustion engines. Ozone injury, or bronzing (Fig. 171), appears on the upper leaf surface first as small, water-soaked or necrotic lesions that may coalesce and become bronze or reddish brown, resembling sunscald injury. Premature senescence and defoliation may then occur. The severity of plant damage is affected by ozone concentration, cultivar sensitivity, leaf age, light (cloud cover), temperature, humidity, soil moisture and texture, and plant nutrition.

PAN forms when oxides of nitrogen interact photochemically with hydrocarbons produced from the incomplete combustion of petroleum products. PAN damage appears on the lower leaf surface initially as water-soaked, shiny or silvery patches that eventually become bronzed.

Sulfur dioxide forms during the combustion of fossil fuels and can act directly as an air pollutant or combine with water to form sulfuric acid. Sulfur dioxide injury may appear on the upper or lower leaf surface as a dull, dark green, water-soaked area that becomes necrotic or bleached. Sulfur dioxide injury is generally more serious on younger than on older leaves, especially when temperature, soil moisture level, and relative humidity are high.

There are other air pollutants that can damage beans, but generally they are not as common as ozone, PAN, or sulfur dioxide. Hydrogen fluoride may cause young leaf tips and margins to become necrotic and curl downward. These symptoms may be severe near sources of hydrogen fluoride, such as aluminum smelters, phosphate fertilizer operations, or other chemical plants. Chlorine gas can induce dark green spots or flecks on the upper leaf surface that later become light tan or brown. Symptoms of injury caused by hydrochloric acid appear as yellow-brown, brown, red, or nearly black flecks or spots surrounded by a cream or white border on margins or interveinal tissue on the upper leaf surface. Hydrochloric acid may also cause a glazing on the lower leaf surface that resembles PAN damage. Nitrogen oxide and nitrogen dioxide can cause chlorotic or bleached symptoms on the upper leaf surface. These symptoms may extend to the lower leaf surface and resemble damage caused by sulfur dioxide. Necrotic lesions induced by nitrogen oxide or nitrogen dioxide may fall out from the leaf, leaving a shot-hole appearance.

The level of air pollution damage by ozone has been reduced on various crops by applying antioxidants. Other management measures may include the development of cultivars less sensitive to damage by air pollutants.

Selected References

Anonymous. 1978. Diagnosing Vegetation Injury Caused by Air Pollution. Publ. 450/3-78-005. U.S. Environmental Protection Agency, Washington, DC.

Katterman, F. 1990. Environmental Injury to Plants. Academic Press, New York.

Manning, W. J., and Feder, W. A. 1980. Biomonitoring Air Pollutants with Plants. Applied Science Publ. Ltd., London, England.

Schwartz, H. F. 1989. Additional problems. Pages 605-616 in: Bean Production Problems in the Tropics, 2nd ed. H. F. Schwartz and M. A. Pastor-Corrales, eds. Centro Internacional de Agricultura Tropical (CIAT), Cali, Colombia.

Zaumeyer, W. J., and Thomas, H. R. 1957. A monographic study of bean diseases and methods for their control. U.S. Dep. Agric. Tech. Bull. 868.

(Prepared by H. F. Schwartz;
Revised by H. F. Schwartz and R. L. Forster)

Baldheads

Baldheads are plants with broken or dead growing points caused by mechanical damage (Fig. 172) during harvest or handling in storage. These plants may develop shoots from the axillary buds at the cotyledonary nodes, but they produce few pods and seeds. A similar injury, called snakehead, can be

Fig. 171. Ozone-induced bronzing of leaves. (Courtesy R. Hall)

caused by the seedcorn maggot; its presence is favored by high levels of organic matter in the soil and it feeds on the flat inner surfaces of the cotyledons. It leaves a feeding track on those inner surfaces and a damaged growing point. Infection by bacterial pathogens can also damage or kill the growing point of a germinating seedling.

Selected References

Schwartz, H. F. 1989. Additional problems. Pages 605-616 in: Bean Production Problems in the Tropics, 2nd ed. H. F. Schwartz and M. A. Pastor-Corrales, eds. Centro Internacional de Agricultura Tropical (CIAT), Cali, Colombia.

Zaumeyer, W. J., and Thomas, H. R. 1957. A monographic study of bean diseases and methods for their control. U.S. Dep. Agric. Tech. Bull. 868.

(Prepared by H. F. Schwartz;
Revised by H. F. Schwartz and R. L. Forster)

Genetic Abnormalities

Beans occasionally exhibit physiological and genetic abnormalities that may be confused with symptoms induced by plant pathogens or abiotic factors. Albino seedlings (Fig. 173) may occur but usually die within a few days because they lack chlorophyll. Leaf variegations (chimeras) may appear as mosaic patterns of green, yellow, and white tissue (Fig. 174) and can cause an abnormal development of the plant and pods. Individual leaves or branches may be affected or the entire plant may express variegations. Environmental stresses, such as low temperatures (below 16°C), that persist during germination can increase the frequency of leaf variegations. General plant chlorosis and pseudomosaic symptoms can be heritable traits (Fig. 175). Small chlorotic spots (yellow spot) may appear on primary and trifoliolate leaves of certain cultivars that otherwise develop normally, and the trait is heritable.

A heritable seedling wilt, not caused by root rot, has been reported. Primary leaves become pale and bronzed, curl slightly, and senesce, resulting in plant death. This problem in breeding nurseries has been attributed to an expression of genetic incompatibility between the parents used in crossing. Crippled or abnormal plant development can occur and may be caused by a genetic abnormality. Internal necrosis is also a heritable trait that produces brown necrotic spots on the flat surface of cotyledons. Seed coat rupture may take place in certain cultivars and appears to be heritable (see Seed Coat Rupture).

Selected References

Schwartz, H. F. 1989. Additional problems. Pages 605-616 in: Bean Production Problems in the Tropics, 2nd ed. H. F. Schwartz and

Fig. 174. Genetic chimera producing yellow and green mosaic in leaves. (Courtesy R. L. Forster)

Fig. 172. Baldhead, caused by mechanical damage to the growing point of the seedling shoots. (Courtesy H. F. Schwartz)

Fig. 173. Albino seedling (right), caused by genetic deficiency in chlorophyll production. Healthy plant is on the left. (Courtesy H. F. Schwartz)

Fig. 175. Heritable pseudomosaic. (Courtesy R. Hall)

M. A. Pastor-Corrales, eds. Centro Internacional de Agricultura Tropical (CIAT), Cali, Colombia.

Zaumeyer, W. J., and Thomas, H. R. 1957. A monographic study of bean diseases and methods for their control. U.S. Dep. Agric. Tech. Bull. 868.

(Prepared by H. F. Schwartz;
Revised by H. F. Schwartz and R. L. Forster)

Green Spot

Green spot describes a condition in which bean plants fail to develop and mature normally. It is usually not detected until beans have begun to senesce, at which time green spots representing dozens or hundreds of adjacent plants become evident in the field (Fig. 176). Affected plants are generally immature, often in full bloom, and appear similar to what might be expected when pods are continually removed from plants as they form. In some cases, pods do form, but they appear puffy and are referred to as puff pods, wind pods, or tough pods. Such pods are similar in size to normal pods and contain seeds that are undeveloped or seeds that are smaller than normal with lumpy, deformed cotyledons (Fig. 177). Deformed seeds, if they germinate and grow, produce normal plants and seeds. Very little research has been done on green spot in the western United States, and the cause remains unknown. In general, white-seeded beans seem to be more affected than are colored beans, and snap beans seem to be affected more than are dry

edible beans. The green spot areas reoccur in the same general field location or locations and appear to be about the same size as previous occurrences.

(Prepared by H. F. Schwartz;
Revised by H. F. Schwartz and R. L. Forster)

Hail Injury

Symptoms of hail injury may include tattered leaves, broken or crushed stems and branches, and long, whitish, bruised areas on stems, branches, petioles, and leaves (Fig. 178). Damage depends on the intensity and duration of the storm and the size of the hailstones, as well as the plant type and its stage of development. Large portions of leaflets are commonly broken off and scattered on the ground. Pods may also be bruised, shredded, crushed, or broken. Developing seeds may be bruised, become discolored, and abort (Fig. 179). Severe hail can reduce stands and delay crop maturity.

If hail strikes the crop during the first 35 days of development, the extent of damage to the first three nodes should be

Fig. 178. Plants damaged by hail. (Courtesy H. F. Schwartz)

Fig. 176. Green spot symptoms in a field. (Courtesy R. L. Forster)

Fig. 177. Seed deformation associated with green spot-affected plants. (Courtesy R. L. Forster)

Fig. 179. Pods damaged by hail. (Courtesy R. L. Forster)

determined and stand counts should be taken to determine whether replanting is necessary. Determinate (bush) types are likely to suffer greater losses when hail occurs during the first 40–45 days of development than are indeterminate (vining) types, because they proceed directly from vegetative into reproductive growth when the first flower appears. Regardless of type, plants have more time to recover and may have less yield reduction when injured earlier in the season.

Hail accompanied by wind and rain may also injure plant tissue and cause water-soaking of large areas of injured tissue. Pathogenic bacteria may easily enter and multiply in this tissue and can be spread through the field by wind and rain.

Selected References

Schwartz, H. F. 1989. Additional problems. Pages 605-616 in: Bean Production Problems in the Tropics, 2nd ed. H. F. Schwartz and M. A. Pastor-Corrales, eds. Centro Internacional de Agricultura Tropical (CIAT), Cali, Colombia.
Zaumeyer, W. J., and Thomas, H. R. 1957. A monographic study of bean diseases and methods for their control. U.S. Dep. Agric. Tech. Bull. 868.

(Prepared by H. F. Schwartz;
Revised by H. F. Schwartz and R. L. Forster)

Lightning Injury

Occasionally, an area of dead and injured plants occurs in the middle of a bean field, and all attempts to diagnose the cause of the problem fail. This problem may be caused by lightning. The roughly circular affected area contains yellow to light brown plants that die within a few days. The stem pith of these plants may be brown to black.

Selected References

Schwartz, H. F. 1989. Additional problems. Pages 605-616 in: Bean Production Problems in the Tropics, 2nd ed. H. F. Schwartz and M. A. Pastor-Corrales, eds. Centro Internacional de Agricultura Tropical (CIAT), Cali, Colombia.
Zaumeyer, W. J., and Thomas, H. R. 1957. A monographic study of bean diseases and methods for their control. U.S. Dep. Agric. Tech. Bull. 868.

(Prepared by H. F. Schwartz;
Revised by H. F. Schwartz and R. L. Forster)

Moisture Stress

Plants may be subjected to high- or low-moisture-level stresses that can influence physiological processes, plant development (Fig. 180), and susceptibility to plant pathogens. In addition to predisposing plants to Fusarium root rot, low soil moisture levels can damage plants in several ways, including the accumulation of toxic concentrations of ions, such as magnesium and boron; closure of stomates; restricted uptake of carbon dioxide; and temporary or permanent wilting.

In contrast, high soil moisture levels and flooding may leach important nutrients required for normal plant development, reduce oxygen content, induce general plant chlorosis (Fig. 181) and stunting, increase levels of toxic by-products from anaerobic metabolism, and predispose plants to various root rots. Combined with high temperatures, excess soil moisture may increase the rate of respiration.

High soil moisture levels coupled with warm soil and low air temperatures may induce intumescence (edema) in cultivars that have abundant foliage and pods that are not directly exposed to the sun. Intumescence is seen on leaves or pods as raised, white to light green spots that occur as a result of a swelling of cells. The spots generally burst, causing small brown spots if favorable conditions persist.

Selected References

Schwartz, H. F. 1989. Additional problems. Pages 605-616 in: Bean Production Problems in the Tropics, 2nd ed. H. F. Schwartz and M. A. Pastor-Corrales, eds. Centro Internacional de Agricultura Tropical (CIAT), Cali, Colombia.
Zaumeyer, W. J., and Thomas, H. R. 1957. A monographic study of bean diseases and methods for their control. U.S. Dep. Agric. Tech. Bull. 868.

(Prepared by H. F. Schwartz;
Revised by H. F. Schwartz and R. L. Forster)

Pesticide Injury

If chemicals are not applied according to manufacturers' recommendations, beans may be damaged during the growing season, especially during germination and seedling development (Figs. 182–184). Even if herbicides are applied correctly, low soil temperatures at seeding retard germination and emergence, which may cause seedlings to remain in contact with preplant-incorporated herbicides for excessively long periods.

Fig. 180. Plant stunting, leaf cupping, and chlorosis caused by low-moisture-level stress. (Courtesy H. F. Schwartz)

Fig. 181. Plant chlorosis caused by high-moisture-level stress. (Courtesy H. F. Schwartz)

This condition stresses the seedling (e.g., root stunting, nutrient leakage, and pruning) and may predispose it to damping-off. High concentrations of various chemicals or fertilizer may be placed too close to seeds, creating toxicity and salt problems if the chemicals do not dissolve and leach rapidly through the root zone. Pesticide drift is common in large production areas intercropped with cereals and can produce distinctive necrotic or morphological symptoms (leaf cupping and distortion) on affected leaves or plant parts (Figs. 185 and 186). Surfactants and crop oils used with pesticides, especially postemergent herbicides, can cause phytotoxic symptoms on foliage.

Root injury by herbicides and other pesticides applied during the current season or previously may be increased by high-soil-moisture-level and low-temperature stresses, deep planting, soil compaction, and mechanically damaged seeds. Chemically damaged roots are often predisposed to subsequent infection, and plants may suffer greater yield loss by root rot pathogens.

Selected References

Anonymous. 1978. Diagnosing Vegetation Injury Caused by Air Pollution. Publ. 450/3-78-005. U.S. Environmental Protection Agency, Washington, DC.

Flor, C. A., and Thung, M. T. 1989. Nutritional disorders. Pages 571-604 in: Bean Production Problems in the Tropics, 2nd ed. H. F. Schwartz and M. A. Pastor-Corrales, eds. Centro Internacional de Agricultura Tropical (CIAT), Cali, Colombia.

Katterman, F. 1990. Environmental Injury to Plants. Academic Press, New York.

Manning, W. J., and Feder, W. A. 1980. Biomonitoring Air Pollutants with Plants. Applied Science Publ. Ltd., London, England.

Schwartz, H. F. 1989. Additional problems. Pages 605-616 in: Bean Production Problems in the Tropics, 2nd ed. H. F. Schwartz and M. A. Pastor-Corrales, eds. Centro Internacional de Agricultura Tropical (CIAT), Cali, Colombia.

Zaumeyer, W. J., and Thomas, H. R. 1957. A monographic study of bean diseases and methods for their control. U.S. Dep. Agric. Tech. Bull. 868.

(Prepared by H. F. Schwartz;
Revised by H. F. Schwartz and R. L. Forster)

Fig. 184. Injury caused by 2,4-D and dicamba. (Courtesy R. L. Forster)

Fig. 182. Injury caused by atrazine. (Courtesy A. F. Sherf)

Fig. 185. Injury caused by drift of paraquat. (Courtesy H. F. Schwartz)

Fig. 183. Injury caused by dicamba. (Courtesy H. F. Schwartz)

Fig. 186. Foliar necrosis caused by quizalofop-P-ethyl plus crop oil. (Courtesy H. F. Schwartz)

pH

Beans grow best at a soil pH of 6.0–7.2. Plant growth declines markedly below pH 5.0 and above pH 8.0. The pH of the soil influences bean growth to a large extent by affecting the supply of minerals. Thus, the symptoms of high (Fig. 187) or low pH are generally those of nutrient deficiencies or toxicities. Above pH 7.2, deficiencies of zinc, manganese, or iron may occur. Growth responses to the addition of these elements or to a reduction of soil pH may be expected if soil tests for the elements indicate low levels. In acidic soils, sulfur and phosphorus deficiencies and aluminum and manganese toxicities are more likely.

Selected References

Anonymous. 1978. Diagnosing Vegetation Injury Caused by Air Pollution. Publ. 450/3-78-005. U.S. Environmental Protection Agency, Washington, DC.

Flor, C. A., and Thung, M. T. 1989. Nutritional disorders. Pages 571-604 in: Bean Production Problems in the Tropics, 2nd ed. H. F. Schwartz and M. A. Pastor-Corrales, eds. Centro Internacional de Agricultura Tropical (CIAT), Cali, Colombia.

Katterman, F. 1990. Environmental Injury to Plants. Academic Press, New York.

Manning, W. J., and Feder, W. A. 1980. Biomonitoring Air Pollutants with Plants. Applied Science Publ. Ltd., London, England.

Schwartz, H. F. 1989. Additional problems. Pages 605-616 in: Bean Production Problems in the Tropics, 2nd ed. H. F. Schwartz and M. A. Pastor-Corrales, eds. Centro Internacional de Agricultura Tropical (CIAT), Cali, Colombia.

Fig. 187. Leaf scorch caused by high soil pH (>8). (Courtesy H. F. Schwartz)

Fig. 188. Sunscald injury on pods (three pods on the right). Healthy pod is on the left. (Courtesy H. F. Schwartz)

Zaumeyer, W. J., and Thomas, H. R. 1957. A monographic study of bean diseases and methods for their control. U.S. Dep. Agric. Tech. Bull. 868.

(Prepared by H. F. Schwartz;
Revised by H. F. Schwartz and R. L. Forster)

Sunscald

Sunscald of bean leaves, stems, branches, and pods may occur during periods of intense sunlight, especially following conditions of high humidity and cloud cover. High temperatures may also induce sunscald damage. Symptoms appear as small, water-soaked spots on the exposed side of the plant. The spots become reddish or brown and may coalesce to form large necrotic or discolored lesions on affected plant organs (Fig. 188). Symptoms may resemble those caused by the tropical spider mite and air pollutants.

Bean development can also be influenced by light intensity, quality, and duration. High light intensity can scorch or burn leaves and pods (russet), cause flower and pod abortion, and increase damage caused by chemical spray droplets or air pollution.

Selected References

Anonymous. 1978. Diagnosing Vegetation Injury Caused by Air Pollution. Publ. 450/3-78-005. U.S. Environmental Protection Agency, Washington, DC.

Flor, C. A., and Thung, M. T. 1989. Nutritional disorders. Pages 571-604 in: Bean Production Problems in the Tropics, 2nd ed. H. F. Schwartz and M. A. Pastor-Corrales, eds. Centro Internacional de Agricultura Tropical (CIAT), Cali, Colombia.

Katterman, F. 1990. Environmental Injury to Plants. Academic Press, New York.

Manning, W. J., and Feder, W. A. 1980. Biomonitoring Air Pollutants with Plants. Applied Science Publ. Ltd., London, England.

Schwartz, H. F. 1989. Additional problems. Pages 605-616 in: Bean Production Problems in the Tropics, 2nd ed. H. F. Schwartz and M. A. Pastor-Corrales, eds. Centro Internacional de Agricultura Tropical (CIAT), Cali, Colombia.

Zaumeyer, W. J., and Thomas, H. R. 1957. A monographic study of bean diseases and methods for their control. U.S. Dep. Agric. Tech. Bull. 868.

(Prepared by H. F. Schwartz;
Revised by H. F. Schwartz and R. L. Forster)

Temperature Stress

Beans are also affected by soil and air temperatures, and sudden changes may influence the plant's ability to absorb soil moisture. Low temperatures may produce chilling or frost damage, apparent as dark, water-soaked areas on wilted leaves or plants (Fig. 189), or they may stunt general plant development if these low temperatures persist for an extended period. Low soil temperatures at planting delay germination and result in reduced stand, especially where poor-quality seeds have been used. The optimum temperature for bean seed germination is 20°C. When soils are cool and wet, the potential for root rot problems and injury from preplant, soil-incorporated herbicides increases.

High temperatures may induce flower abortion, increase the rate of evapotranspiration, and cause leaf wilting or necrosis if there is an insufficient supply of soil moisture or if root growth is limited (e.g., by root rot, mechanical damage, or soil compaction). High temperatures and winds may compound the stresses induced by low soil moisture levels by physically inducing soil aggregation, cracks, and subsequent root damage. Seedlings may develop stem lesions at the soil line if the soil surface becomes too hot.

Selected References

Anonymous. 1978. Diagnosing Vegetation Injury Caused by Air Pollution. Publ. 450/3-78-005. U.S. Environmental Protection Agency, Washington, DC.

Flor, C. A., and Thung, M. T. 1989. Nutritional disorders. Pages 571-604 in: Bean Production Problems in the Tropics, 2nd ed. H. F. Schwartz and M. A. Pastor-Corrales, eds. Centro Internacional de Agricultura Tropical (CIAT), Cali, Colombia.

Katterman, F. 1990. Environmental Injury to Plants. Academic Press, New York.

Manning, W. J., and Feder, W. A. 1980. Biomonitoring Air Pollutants with Plants. Applied Science Publ. Ltd., London, England.

Schwartz, H. F. 1989. Additional problems. Pages 605-616 in: Bean Production Problems in the Tropics, 2nd ed. H. F. Schwartz and M. A. Pastor-Corrales, eds. Centro Internacional de Agricultura Tropical (CIAT), Cali, Colombia.

Fig. 189. Freeze-damaged (left) and nondamaged (right) pods and seeds. (Courtesy H. F. Schwartz)

Fig. 190. Injury to leaves caused by windburn and rubbing. (Courtesy R. L. Forster)

Zaumeyer, W. J., and Thomas, H. R. 1957. A monographic study of bean diseases and methods for their control. U.S. Dep. Agric. Tech. Bull. 868.

(Prepared by H. F. Schwartz;
Revised by H. F. Schwartz and R. L. Forster)

Wind and Sand Damage

Wind speed and direction can affect plant development. Evapotranspiration rates may be increased by consistent winds and thereby aggravate moisture stress caused by low soil moisture levels. Hot, dry winds may cause hydathodes and cells around margins of expanding leaves to die, resulting in a fine brown line of dead cells on the leaf margin and the leaf-cupping several days later as the cells in the rest of the leaf continue to enlarge. Violent wind movement may damage roots and predispose them to subsequent root rot problems, break stems and branches, and cause plant lodging, especially if the soil moisture level is high. Leaves may be abraded, torn, or shredded.

Beans can also be damaged by the abrasive action of wind and airborne soil particles (Fig. 190). Buds, blossoms, and young pods may be lost from the combined action of such factors. Strong winds may scatter beans that are cut and in windrows ready for threshing, causing loss in harvested seeds.

Selected References

Schwartz, H. F. 1989. Additional problems. Pages 605-616 in: Bean Production Problems in the Tropics, 2nd ed. H. F. Schwartz and M. A. Pastor-Corrales, eds. Centro Internacional de Agricultura Tropical (CIAT), Cali, Colombia.

Zaumeyer, W. J., and Thomas, H. R. 1957. A monographic study of bean diseases and methods for their control. U.S. Dep. Agric. Tech. Bull. 868.

(Prepared by H. F. Schwartz;
Revised by H. F. Schwartz and R. L. Forster)

Mineral Deficiencies and Toxicities

Beans require fertile soils for good growth. In Central and South America, deficiencies of phosphorus, nitrogen, and certain micronutrients are common problems in many soils. In acidic soils, toxicity caused by excess levels of aluminum and manganese may develop. In North America, deficiencies in manganese, zinc, and iron have been reported. Little information is available from Africa and western Asia.

Aluminum

Plants affected by aluminum toxicity are small and have poorly developed root systems, numerous adventitious roots near the soil surface, and chlorotic lower leaves with necrotic margins. Aluminum toxicity results from high levels of aluminum in the soil solution, generally when aluminum represents 10% or more of the total exchange capacity (1–2 meq/100 ml of soil) of aluminum, calcium, magnesium, and potassium in the soil. The problem is common in acidic soils (below pH 5.0), which are widespread in Latin America, and is strongly related to phosphorus and calcium deficiencies.

Aluminum toxicity is usually corrected by the addition of lime (1–5 t/ha) to the soil to raise the pH and adjust the proportions of calcium and, if dolomitic lime is used, magnesium. Cultivars of dry beans vary in their sensitivity to aluminum toxicity, and some are tolerant to moderate levels of aluminum.

Boron

Boron deficiency can occur in coarse-textured soils having low organic matter content and high levels of aluminum and iron hydroxide, in alluvial soils with high pH and low total boron content, and in neutral or alkaline soils subjected to dryness and high light intensity. Liming the soil may induce boron deficiency. The first symptom of boron deficiency is the death of the main growing tip. Lateral buds produce many small branches, but their terminal buds die. Primary leaves thicken and become deformed and leathery. Trifoliolate leaves may form only one or two deformed leaflets, and petioles become brittle. Interveinal chlorosis occurs on all leaves. Stems are swollen near nodes. Flowers and pods either abort or do not form, and the root system is poorly developed. Longitudinal cracks appear near the base of the stem. Symptoms are intensified by low soil moisture levels. Cultivars of dry beans vary in their sensitivity to boron deficiency, and some are tolerant. Boron deficiency may be corrected by applying boric acid to the soil or soluble formulations of boron to the foliage.

Boron toxicity causes yellowing and necrosis of the margins of primary leaves shortly after emergence (Fig. 191), as well as of older leaves. Toxicity symptoms may appear when the boron content of the soil exceeds 5 ppm. The toxicity generally occurs after nonuniform application of fertilizer or when the fertilizer is band-applied too close to the seeds, especially during dry weather. Beans following crops receiving high levels of boron fertilizer (e.g., turnips) may show symptoms of boron toxicity. The problem may generally be avoided by careful use of fertilizers and crop sequences.

Fig. 191. Yellowing, necrosis, and stunting caused by excess boron. From left to right, 1, 6, 9, and 12 mg of boron per liter, respectively. (Courtesy U.S. Salinity Laboratory, Riverside, CA)

Fig. 192. Chlorosis of younger leaves caused by iron leaching from plants exposed to excess water. (Courtesy H. F. Schwartz)

Calcium

Acidic soils with a pH between 4 and 5.5 normally have low levels of calcium and magnesium. Calcium deficiency often occurs together with aluminum and manganese toxicities. Symptoms include loss of turgor, death of growing points, dark green older leaves, yellowing of newer leaves, and sometimes hypocotyl collapse. Older leaves eventually scorch and senesce. Pods may also be soft and yellowed, and seeds may fail to develop. Plant height and dry matter production are often reduced. Calcium deficiency may be corrected by liming and application of superphosphate fertilizer.

Copper

Copper deficiency rarely affects beans but can occur in organic, sandy, and overlimed acidic soils. Plants are stunted and have short internodes. Young leaves are pale to gray or blue-green. Irregular necrotic areas or blotches appear close to veins near the base of the leaflet. Scorching may develop on one side of the leaflet, which then wilts and senesces. Growing tips may die back, and flowering can be suppressed. Copper deficiency can be corrected by applying copper sulfate (5–10 kg/ha). Minor deficiencies can be corrected by foliar applications of copper sulfate or copper chelate.

Iron

Iron deficiency can occur in calcareous soils containing free calcium carbonate, in alkaline soils, or in acidic soils that have been overlimed. Excessive phosphate may precipitate the available iron as insoluble iron phosphate. A temporary deficiency (less than 48 h) may occur on younger leaves of bean plants grown in alkaline soils that have been saturated with moisture from rainfall or irrigation, especially if daytime temperatures are low.

Symptoms of iron deficiency appear in young leaves (Fig. 192), which develop interveinal yellowing while the veins remain green (Fig. 193). In severe cases, leaves may become almost white. Profuse, irregular necrosis may develop on severely chlorotic leaves. Fully expanded leaves curve downward and leaf tips may wilt. Young, unexpanded leaflets may senesce. Iron deficiency can be corrected by soil application of iron chelates or foliar application of iron salts (e.g., 0.5% $FeSO_4$). Some cultivars are less sensitive to low levels of iron.

Magnesium

Magnesium deficiency generally occurs in acidic soils with low base levels and on volcanic ash soils with low levels of potassium and calcium. Symptoms include interveinal chlorosis and rusty speckling to necrosis of the upper surfaces of lower leaves. These interveinal spots may be 0.5 mm in diameter, and they are angular and slightly sunken. Magnesium deficiency is normally corrected by the application of dolomitic lime.

Fig. 193. Interveinal chlorosis and general chlorosis of younger leaves caused by iron deficiency. (Courtesy H. F. Schwartz, from the files of CIAT)

Manganese

Manganese deficiency can occur in alkaline, organic, poorly drained, or overlimed acidic soils. Symptoms appear as interveinal chlorosis and fine speckling on younger leaves (Figs. 194 and 195). These leaves may also appear pimply when viewed close up. Older leaves are smoother and generally chlorotic. Pods may be yellow and unfilled. Plants are dwarfed. Foliar sprays of manganese salts (e.g., 0.5% $MnSO_4$) can correct the problem.

Manganese toxicity occurs in volcanic soils with a low pH. Poor drainage aggravates the problem. Symptoms may appear as purple-black spots on the stem, petiole, midrib, and veins of leaves, especially the lower leaf surface. The pulvinus region is not discolored. Microscopic examinations reveal that these spots are clumps of secreted material (manganese dioxide) around the basal cells of hairs. Chlorosis may develop between major veins, especially on younger leaves. Affected leaves may cup downward and have necrotic margins. Some cultivars are less sensitive to manganese toxicity than are others. Improved drainage, addition of organic matter, and application of lime may alleviate the problem.

Molybdenum

Molybdenum deficiency can occur in acidic soils, especially if iron and aluminum contents reduce molybdenum solubility. Symptoms resemble those of nitrogen deficiency. The deficiency may be corrected by liming the soil.

Nitrogen

Beans are legumes and can fix nitrogen in the presence of appropriate strains of the genus *Rhizobium* unless cultural, varietal, or inoculation difficulties limit this fixation ability and leave the plant dependent on residual soil nitrogen or applied nitrogen fertilizer. Nitrogen deficiency can occur in all soils and is especially severe in sandy soils with low levels of organic matter. Symptoms appear as a uniformly pale green to yellow discoloration of all leaves except the younger ones. Growth is reduced, few flowers develop, or pods fill poorly. A small amount of nitrogen fertilizer is often recommended at seeding to boost seedlings before nitrogen fixation occurs.

Nitrogen deficiency can be corrected by the application of nitrogen fertilizer and organic matter. The appropriate recommendation of fertilizer depends on factors such as the type of soil and the cropping history of the field. Soil and seeds can be inoculated with appropriate strains of the genus *Rhizobium*, but success is affected by specificity between the *Rhizobium* strain and its host cultivar, interaction with native and less-efficient rhizobia, soil pH, soil temperature, nitrogen and phosphorus contents of soil, pesticide use, and farming practices.

Phosphorus

Phosphorus deficiency can occur in many soils, especially those with a low pH. Symptoms appear initially on upper leaves, which are small and dark green, while older leaves may turn brown and senesce prematurely. Plants are often stunted and have thin stems and shortened internodes. The vegetative period may be prolonged, whereas the flowering phase is delayed and shortened. Often, many flowers abort, and the number of pods and seeds may be reduced. Phosphorus deficiency can be corrected by band application of various rock phosphates or superphosphate fertilizers. Some cultivars are more tolerant of low phosphorus levels than are others.

Potassium

Potassium deficiency may occur in soils with low fertility, high calcium and magnesium contents, or a highly permeable, sandy texture. Symptoms appear first as a marginal chlorosis of older leaves, followed by a marginal scorch between the veins (similar to symptoms of common bacterial blight, but without the water-soaking). The leaf may curl downward, but scorched margins curve upward. Plants may have weak stems, short internodes, and reduced root growth, and they can collapse easily. Potassium deficiency can be corrected by the application of potassium chloride or potassium sulfate. Some cultivars of beans are tolerant because they can efficiently use small amounts of potassium in the soil.

Sodium and Salinity

Beans are very sensitive to the salinity or sodium content of the soil. In general, sodium content becomes a problem when saturation is more than 4% and soil compaction is moderate to severe. Salinity adversely affects beans when the conductivity of the soil solution is more than 2 mmho/cm. Growth reduction, leaf scorch, and plant death may occur when sensitive bean cultivars are planted in saline soils or soils with a high sodium content. Damage may be high during germination and seedling development, and plant stands can be significantly reduced. Salinity of the soil is a difficult problem to solve. Increased

Fig. 194. Early symptoms of interveinal chlorosis on a younger leaflet caused by manganese deficiency. (Courtesy A. W. Saettler, from the files of D. R. Christenson)

Fig. 195. Advanced symptoms of interveinal chlorosis on younger leaves caused by manganese deficiency. (Courtesy A. W. Saettler, from the files of D. R. Christenson)

salinity in the root zone can be avoided by careful management of irrigation water, and excess soil salinity can be corrected by leaching with water and using deep-rooted crops.

Sulfur

Sulfur deficiency seldom occurs in bean production areas anywhere in the world. Symptoms are similar to those of nitrogen deficiency. A pronounced, uniform, yellow chlorosis of the youngest leaves develops and growth ceases. Older leaves may also be chlorotic. Sulfur deficiency can generally be corrected by the application of powdered sulfur (10–20 kg/ha) or fertilizers, such as ammonium sulfate or superphosphate.

Zinc

Zinc deficiency can occur in soils with a high pH or in acidic soils that have received too much lime or phosphorus. The problem can also be aggravated by soil compaction, low levels of organic matter, and excess application of manure or crop residue. Elevated absorption of other nutrients, such as iron, can also induce zinc deficiency. Soon after emergence, younger leaves develop an interveinal chlorosis and become deformed, dwarfed, and crumpled (Figs. 196 and 197). Older leaves may develop necrotic areas in and between veins. Terminal blossoms and pods may abort. If the deficiency is severe, new leaves are white and plants may die. Deficiencies may appear in spots within a field or throughout the entire field. Zinc deficiency can be corrected by soil (6–12 kg/ha) or foliar (0.5% solution) application of zinc sulfate. Some cultivars are more sensitive to low levels of zinc than are others. Small-seeded navy and black bean cultivars, such as Sanilac, Makinac, and T-39, are highly sensitive. Pinto, great northern, pink, and small red market classes are, in general, more resistant to zinc deficiency.

Selected References

Flor, C. A., and Thung, M. T. 1989. Nutritional disorders. Pages 571-604 in: Bean Production Problems in the Tropics, 2nd ed. H. F. Schwartz and M. A. Pastor-Corrales, eds. Centro Internacional de Agricultura Tropical (CIAT), Cali, Colombia.

Hall, R., and Schwartz, H. F. 1993. Common bean. Pages 143-147 in: Nutrient Deficiencies & Toxicities in Crop Plants. W. F. Bennett, ed. American Phytopathological Society, St. Paul, MN.

Scaife, A., and Turner, M. 1983. Phaseolus beans. Pages 63-67 in: Diagnosis of Mineral Disorders in Plants, Vol. 2: Vegetables. Her Majesty's Stationery Office, London.

Zaumeyer, W. J., and Thomas, H. R. 1957. A monographic study of bean diseases and methods for their control. U.S. Dep. Agric. Tech. Bull. 868.

(Prepared by H. F. Schwartz;
Revised by H. F. Schwartz and R. L. Forster)

Seed Quality

Germination and seedling establishment may be affected by the overall quality of the seeds planted. Seed quality has long been a problem in beans, especially with white-seeded classes, such as snap and navy beans. Poor seed quality is expressed by numerous symptoms, such as reduced germination, poor seedling vigor, reduced ability to emerge (especially if there is a soil crust), and various abnormalities of plant parts. Seedlings that are delayed or weak in emergence contribute little to yield and, in the case of snap beans, may function as "weeds" in the crop.

Mechanical Damage

Poor seed quality has been blamed on rough handling during harvest, low (less than 11%) seed moisture levels during harvesting and handling, or both, resulting in increased susceptibility to mechanical damage (MD); on planting in cold, wet soils, resulting in a sudden shock to the seeds during imbibition; on a genetic predisposition to transverse cotyledon cracking (TVC); and on the general association of white seed color with a predisposition to damage. The causes are related, since the level of TVC may be increased in dry seeds, may result from MD, or may result from planting in cold, wet soil. In any case, the cracks prevent the cotyledon from feeding the seedling, as can be seen by the plump appearance in the cotyledon distal to the crack (Figs. 198 and 199). In extreme cases, MD may result in seedlings with broken or dead growing points, also known as baldheads (Fig. 172).

Storage at high moisture content, greater than 15%, or for long periods of time may result in loss of viability or vigor. Standard germination tests give a good indication of viability under normal planting conditions. Soaking a seed sample in a 1% tetrazolium chloride solution for 12 h helps identify potentially low-vigor lots that may give problems under less-than-ideal planting conditions.

Genetic studies of tolerance to MD and TVC in snap beans suggested quantitative inheritance for resistance with narrow-sense heritability in the range of 27 to 52%. In crosses of resistant, colored × susceptible, white seeds, the F_2 means approached that of the resistant parent for colored segregants and of the susceptible white-seed parent for white segregants. Colored seeds cause discoloration of liquor in canned snap beans and of the cut ends of beans in general; so for many market classes, it is simply not an option. Fortunately, there was a small percentage of white segregants with excellent resistance.

Dropping seeds onto a steel plate several times has been used as a selective tool to identify lines with tolerance to MD. Studies on the effect of two, four, and eight drops from 1.2 m onto

Fig. 196. Stunting caused by zinc deficiency. (Courtesy R. L. Forster)

Fig. 197. Interveinal chlorosis and necrosis of leaves caused by zinc deficiency. (Courtesy H. F. Schwartz, from the files of CIAT)

a steel plate showed dramatic differences among 20 lines in tolerance to MD. In a study with 68 lines and cultivars in which the seeds were dropped four times, germination ranged from 60 to 95% in undropped seeds and from 3 to 79% in dropped seeds. The loss in seedling weight was 7% in the most-resistant line and 73% in the most-susceptible line. The susceptible lines had more TVC, more seed coat shattering, and lighter seed coats. Dropping a large F_3 sample repeatedly until only a small percentage of the seeds was undamaged significantly increased tolerance to MD in just one generation.

An alternative procedure was to select for semihard seeds, in which dry seeds imbibed slowly, but seeds with more than 10% moisture imbibed normally or rapidly. This approach emphasizes the importance of permeability of the seed coat: those lines that are less permeable, or semihard seeds, all have excellent seed coats, have no seed defects, and are more MD resistant, as well as being more tolerant of planting in cold, wet soils. Vigorous selection in white-seeded lines for MD tolerance and good seed quality can be successful.

Selected References

Dickson, M. D., and Boettger, M. A. 1976. Factors associated with resistance to mechanical damage to snap beans (*Phaseolus vulgaris* L.). J. Am. Soc. Hortic. Sci. 101:541-544.

Dickson, M. D., and Boettger, M. A. 1977. Inheritance of resistance to mechanical damage and transverse cotyledon cracking in snap beans (*Phaseolus vulgaris* L.). J. Am. Soc. Hortic. Sci. 102:498-501.

Dickson, M. D., and Boettger, M. A. 1982. Heritability of semi-hard seed induced by low moisture in beans (*Phaseolus vulgaris* L.). J. Am. Soc. Hortic. Sci. 107:69-71.

Dickson, M. D., Duczmal, K., and Shannon, S. 1973. Imbibition rate and seed composition as factors affecting transverse cotyledon cracking in bean. J. Am. Soc. Hortic. Sci. 98:509-513.

Fig. 198. Bean seed transverse cotyledon cracking effects. (Courtesy D. M. Webster)

Fig. 199. Transverse cracking of cotyledons. (Courtesy M. H. Dickson)

Korban, S. S., Coyne, D. P., Weihing, J. H., and Hanna, M. A. 1981. Testing methods, variation, morphological and genetic studies of seed-coat cracking in dry beans (*Phaseolus vulgaris* L.). J. Am. Soc. Hortic. Sci. 106:821-828.

Morris, J. L., Campbell, F. W., and Pollard, L. H. 1970. Relation of imbibition and drying in cotyledon cracking in snap beans, *Phaseolus vulgaris* L. J. Am. Soc. Hortic. Sci. 95:541-543.

Pollock, B. M., and Manelo, J. R. 1970. Simulated mechanical damage to garden bean seed during germination. J. Am. Soc. Hortic. Sci. 95:414-417.

Pollock, B. M., Roos, E. E., and Manelo, J. R. 1960. Vigor of garden bean seed and seedlings as influenced by initial seed moisture, substrate oxygen, and imbibition temperature. J. Am. Soc. Hortic. Sci. 94:577-584.

Wijuandi, S., and Copeland, L. O. 1974. Effect of origin, moisture content, maturity, and mechanical damage on seed and seedling vigor of beans. Agron. J. 66:546-548.

(Prepared by M. H. Dickson; Revised by D. Webster)

Seed Coat Rupture

Seed coat rupture (SCR) in beans is also known as splitting cotyledons, fish face, or fish head. The testa ruptures at the apex of the cotyledons during development (Fig. 200). The exposed cotyledons extend beyond the seed coat and appear cone shaped, roughened, and serrated.

SCR is caused by uneven growth of the cotyledons and seed coats. It has been suggested, but not proven, that the seed coat ruptures when large- and small-seeded cultivars are crossed, resulting in a combination of large cotyledons and small seed coats. SCR seeds are susceptible to decay, although under favorable conditions, they germinate as well as do normal seeds. The amount of SCR varies greatly among lines, from 1 to 48% in one report. Well-watered plants produced a higher percentage of SCR than did those under water stress after flowering. The rate of seed development was similar in both types of seeds until SCR occurred, after which the undamaged beans continued to grow and became heavier, while the SCR beans did not. SCR can occur as early as halfway through the process of seed development.

When F_2 lines were classified as susceptible, intermediate, or resistant to SCR, there were year-to-year fluctuations, but the lines bred true. Crosses of susceptible × resistant types resulted in a good fit to a 1:2:1 ratio of susceptible, intermediate, and resistant. SCR is governed by a single, incompletely dominant gene that is sensitive to the environment and shows only 25–50% penetrance. The gene for susceptibility to SCR was designated *Tr* (testa rupture).

Fig. 200. Seed coat rupture in colored (top row) and white-seeded (bottom row) beans. (Courtesy M. H. Dickson)

Selected References

Dickson, M. H. 1969. The inheritance of seed coat rupture in snap beans, *Phaseolus vulgaris* L. Euphytica 18:110-115.
Farooqui, H. M., and McCollum, J. P. 1954. Relation of morphological structure and development to seed coat rupture in beans (*Phaseolus vulgaris* L.). Proc. Am. Soc. Hortic. Sci. 63:333-346.
Zaumeyer, W. J., and Thomas, H. R. 1957. A monographic study of bean diseases and methods for their control. U.S. Dep. Agric. Tech. Bull. 868.

(Prepared by M. H. Dickson; Revised by D. Webster)

Hypocotyl Collar Rot

Certain market classes, especially Romano-type snap beans, are prone to a calcium deficiency known as hypocotyl collar rot, which is most frequently seen in germination tests on nutrient-poor substrates. Young seedlings exhibit a constricted band around the hypocotyl (Fig. 201). If observed in a germination test, the sample is retested using a 0.3–0.6% calcium nitrate solution to moisten the substratum.

Selected Reference

Association of Official Seed Analysts. 1988. Rules for Testing Seeds, Sect. 4.8 (j). The Association, Las Cruces, NM.

(Prepared by M. H. Dickson; Revised by D. Webster)

Fig. 201. Hypocotyl collar rot symptoms. (Courtesy D. M. Webster)

Glossary

C—Celsius or centigrade ($°C = (°F - 32) \times {}^5/_9$)
cm—centimeter (1 cm = 0.01 m)
g—gram (1 g = 0.03527 ounce; 453.6 g = 1 pound)
h—hour
ha—hectare (1 ha = 10,000 m^2 = 2.471 acres)
kg—kilogram (1 kg = 1,000 g = 2.205 pounds)
km—kilometer (1 km = 1,000 m = 0.6214 mile)
m—meter (1 m = 39.37 inches)
meq—milliequivalent
min—minute
ml—milliliter (1 ml = 0.001 liter)
mm—millimeter (1 mm = 0.001 m)
mmho—millimho (unit of electrical conductivity)
μm—micrometer (1 μm = 10^{-6} m)
nm—nanometer (1 nm = 10^{-9} m)
ppm—parts per million
S—Svedberg (unit of rate of sedimentation in a centrifuge)
t—metric ton (1 t = 1,000 kg)

abaxial—directed away from the stem of a plant; pertaining to the lower surface of a leaf
abscise (n. abscission)—to separate from a plant, as leaves, flowers, and fruits do
acervulus (pl. acervuli)—saucer-shaped or cushionlike fungal fruiting body bearing conidiophores, conidia, and sometimes setae
acicular—slender and pointed; needle-shaped
acidic soil—soil with an acidic reaction (less than pH 7)
acropetal—upward from the base to the apex of a shoot
adaxial—directed toward the stem of a plant; pertaining to the upper surface of a leaf
adventitious—arising from other than the usual place, as roots from a stem rather than as branches of a root
aeciospore—binucleate spore produced in an aecium
aecium (pl. aecia, adj. aecial)—the fruiting body of a rust fungus in which the first binucleate spores are produced
aerobic—requiring oxygen to complete the life cycle
aerosol—a suspension of small particles in air (or other gas)
agar—jellylike material derived from algae and used to solidify liquid media; term also applied to the culture medium itself
alkaline—having basic (nonacidic) properties
allantoid—slightly curved with rounded ends; sausagelike in form
allele—one of a group of genes that occurs at a given site on a chromosome
alluvial—deposited by flowing water
amphigenous—developing growth all around or on two sides
ampulla (pl. ampullae)—a swollen, globular cell on which all the conidia develop simultaneously, as in *Botrytis* spp.
anamorph—the asexual form (also called the imperfect state) in the life cycle of a fungus, when asexual spores (such as conidia) or no spores are produced
anastomosis (pl. anastomoses)—fusion, as of hyphal strands, and combination of their contents
annual—a plant that completes its life cycle and dies within 1 year
anterior—head end
anther—the pollen-bearing portion of a stamen
antheridium (pl. antheridia)—male sexual organ found in some fungi
anthracnose—disease caused by acervuli-forming fungi (order Melanconiales) and characterized by sunken lesions and necrosis

antibiotic—chemical toxic to bacteria; bactericide
antibody—a specific protein formed in the blood of warm-blooded animals in response to the injection of an antigen
antigen—any foreign chemical (normally a protein) that induces antibody formation in animals
antioxidant—agent that limits or prevents oxidation
antiserum (pl. antisera)—blood serum containing antibodies
apex (pl. apexes, adj. apical)—tip of a root or shoot, containing the apical meristem
apiculate—ending abruptly in a small, distinct point
aplerotic—pertaining to an oospore that does not fill the oogonium
apothecium (pl. apothecia)—open, cuplike or saucerlike, ascus-bearing fungal fruiting body
appressed—lying close or flat against a surface
appressorium (pl. appressoria)—swollen, flattened portion of a fungal filament that adheres to the surface of a higher plant, thus providing anchorage for invasion by the fungus
arachnoid—like a cobweb
ascocarp—sexual fruiting body (ascus-bearing organ) of an ascomycete
ascomatum (pl. ascomata)—fruiting body bearing ascospores
ascomycete—member of a class of fungi that produce sexual spores (ascospores) endogenously within an ascus
ascospore—sexual spore borne in an ascus
ascus (pl. asci)—saclike cell in which ascospores (typically eight) are produced
aseptate—lacking septa
asexual—vegetative; without sex organs, sex cells, or sexual spores, as the anamorph of a fungus
autoecious—in reference to rust fungi, producing all spore forms on one species of plant
axil—the angle formed by a leaf petiole and the stem
axillary bud—bud that develops in the axil of a leaf (also called a lateral bud)

bacilliform—shaped like a blunt, thick rod
bactericide—agent, especially chemical, that kills bacteria
baldhead—stunting of a bean plant caused by damage to the vegetative apex of the seedling
basidiomycete—member of a class of fungi that forms sexual spores (basidiospores) on a basidium
basidiospore—haploid spore of a basidiomycete
basidium (pl. basidia, adj. basidial)—short, club-shaped fungus cell on which basidiospores are produced
basipetal—downward from the apex toward the base of a shoot; referring to development in the direction of the base so that the apical part is oldest
bicellular—two-celled
binucleate—containing two nuclei
biological management—disease or pest management by microorganisms and other natural components of the environment
biotype—distinctive biological form within a species
blastic—formed by budding: there is marked enlargement of a recognizable conidium initial before the initial is delimited by a septum
blight—sudden, severe, and extensive spotting, discoloration, wilting, or destruction of leaves, flowers, stems, or entire plants, usually affecting young, growing tissues; in disease names, may be coupled with the name of the affected part of the plant (e.g., leaf blight), the kind of causal organism (e.g., bacterial blight), or a distinctive symptom (e.g., web blight)

bloom—flowering

botryose—shaped like a bunch of grapes

breeding line—plant strain used in a plant breeding program and usually containing one or more desirable agronomic or breeding characteristics

bursa—extension or flap of cuticle at the side of the male nematode sex organ, used for clasping or orienting during mating

bush-type bean—bean plant with determinate growth

calcareous—rich in calcium carbonate (lime)

canker—necrotic, localized diseased area

canopy—a mass of leaf-bearing shoots, measured in height, width, or distribution

capsid—protein coat of a virus

carbohydrate—any of various chemical compounds of carbon, hydrogen, and oxygen, such as sugars, starches, and cellulose

catena (adj. catenulate)—chain, e.g., of spores

causal agent—organism or agent that produces a given disease

chimera—part or organ consisting of genetically different tissues; in beans, manifested often as leaf variegation

chlamydospore—thick-walled or double-walled asexual resting spore formed by modification of a segment of a hypha or spore

chlorophyll—green pigment of plants that absorbs light energy and makes it effective in photosynthesis

chloroplast—disklike structure containing chlorophyll in the cells of green plants

chlorosis (adj. chlorotic)—abnormal plant color of light green or yellow resulting from incomplete formation or destruction of chlorophyll

chromosome—stringlike or beadlike structure in the nucleus that contains genes

clamp connection—in basidiomycete fungi, a branch uniting adjacent cells of a single hypha

clavate—club-shaped

colonize—to permeate plant tissue with growth of a pathogen

colony—growth of a microorganism in mass, especially as a pure culture in the laboratory

conidiogenous—producing conidia

conidiophore—specialized fungal hypha on which conidia are produced

conidium (pl. conidia)—asexual spore formed by abstriction and detachment of part of a hyphal cell at the end of a conidiophore and germinating by a germ tube

corolla—petals

cortex (adj. cortical)—region of parenchyma tissue between the epidermis and the phloem in stems and roots; region of thin-walled, nonpigmented cells beneath the rind of a sclerotium

cotyledon—seed leaf of a germinating plant

cross-fertilization—mating between separate individuals or individuals of different kinds

crustose—formed in a hard, thin layer, like a crust

cultivar (abbr. cv.)—a cultivated plant variety or cultural selection

culture—artificial growth and propagation of organisms on nutrient media or living plants

cupulate—cuplike in form

cuticle (adj. cuticular)—the water-repellent, waxy covering (cutin) of epidermal cells of plant parts, such as leaves, stems, and fruits; the outer sheath or membrane of nematodes

cutin—see cuticle

cylindric—shaped like a cylinder

cyst—resting spore of some fungi

cytoplasm—inner substance of a cell exclusive of the nucleus

damping-off—death of a seedling before or shortly after emergence; it is common to distinguish between preemergence damping-off and postemergence damping-off

defoliation—loss of leaves from a plant

dehiscent—opening by breaking into parts

denticle—fine projection bearing a spore

determinate—ceasing vegetative growth when the first flower forms, as beans with a bush habit

diagnostic (n. diagnosis)—distinctive, as of a distinguishing characteristic serving to identify or determine the presence of a disease or other condition

diapause—a period of dormancy in the life cycle of animals, such as insects

dichotomous—branching, often successively, into two more or less equal arms

dicotyledon—plant with two cotyledons, e.g., bean

dieback (v. die back)—progressive death of shoots, leaves, or roots, beginning at the tips

differential cultivar—plant cultivar that develops a distinctive symptom or a distinctively severe symptom that can be used to assist in identification of a pathotype

differentiation—the physiological and morphological changes that occur in a cell, tissue, or organ during development from a juvenile state to a mature state

dikaryotic—containing two genetically different nuclei

dilution end point—degree of dilution of a virus solution at which it loses infectivity

diploid—having two complete sets of chromosomes (2n chromosomes)

discoid—resembling a disk

disease—abnormal functioning of an organism

disease cycle—succession of events in the life cycle of a pathogen and its relationships with its host and the environment that contribute to the process of disease: includes survival, dissemination, germination, infection, growth, and sporulation by the pathogen; cycle of development of the host plant; activity of vectors; and the interaction of these with weather and other environmental factors

dissemination—spread of infectious material (inoculum) from diseased to healthy plants

distal—far from the point of attachment or origin, as opposed to proximal

DNA—deoxyribonucleic acid

dolipore septum—septum found in basidiomycetes and characterized by special swellings and membranes in association with the septal pore

domesticate—domesticated form of a plant

dominant—pertaining to the form of a gene that is expressed phenotypically when it is present in the plant; a dominant character is a phenotypic characteristic of the plant managed by a dominant gene (see recessive)

dormancy—nongrowing condition of a plant, caused by internal factors or environmental factors

echinulate—spiny

ecology—study of the behavior of organisms in their environment

edema—burst intumescent spots (see intumescence)

effuse—spread out flat, without definite form

electrophoresis—separation of molecules according to their electric charge

ELISA—see enzyme-linked immunosorbent assay

ellipsoid—elliptical in plane section

embryo—young plant in seed, consisting of the plumule, radicle, and cotyledons

emergence—growth of the seedling shoot through the surface of the soil

enation—epidermal outgrowth

encapsidate—to cover virus nucleic acid with a protein coat

encyst—to form a cyst

endemic—native to or peculiar to a locality or region

endocellular—inside the cell

endoconidium (pl. endoconidia)—conidium produced from within the conidiophore

endodermis—layer of cells within the root, between the vascular tissue and the cortex

endoparasite—parasite living within its host

endosperm—nutritive tissue formed within the embryo sac of seed plants

enzyme—protein that catalyzes a specific biochemical reaction

enzyme-linked immunosorbent assay (ELISA)—a serological test in which the sensitivity of the antibody–antigen reaction is increased by attaching an enzyme to one of the two reactants

epidemic—general and serious outbreak of disease

epidemiology (adj. epidemiologic)—the study of factors influencing the initiation, development, and spread of infectious disease; the study of disease in populations of plants

epidermis (adj. epidermal)—outermost layer of cells on plant parts

epigeal germination—germination characterized by the production of cotyledons raised above the soil level

epinasty—downward curvature of a leaf, leaf part, or stem

epiphyte—organism growing on a plant surface

epistatic gene—gene that suppresses the effect of a nonallelic gene

eradicant—chemical used to eliminate a pathogen from a host or an environment

erumpent—breaking out or erupting through the surface

exogenous—originating from the outside

exudates—substance that is excreted or discharged; ooze

f. sp.—*see* forma specialis

facultative—capable of changing life-style, e.g., from saprophytic to parasitic or the reverse

fallow—cultivated land kept free from a crop or weeds during the normal growing season

fasciation—malformation in shoots or floral organs manifested as enlargements, flattening, and sometimes curving as if several parts were fused

fascicle—loose bundle of conidiophores

fastidious—in reference to prokaryotic organisms, having special growth and nutritional requirements

feeder root—fine root that absorbs water and dissolved nutrients

field capacity—water content of soil after it has been wetted and allowed to drain

filament (adj. filamentous)—thin, flexible, threadlike structure; the stalklike, anther-bearing portion of a stamen

filiform—long, needlelike

flaccid—wilted, lacking turgor

flagellum—flexuous appendage providing motive force for a zoospore

forma specialis (abbr. f. sp.) (pl. formae speciales)—a special form of a pathogen species distinguished from other formae speciales by its host range, usually consisting of distinct species of plants, or by the kinds of symptoms produced (*see* race)

fruiting body—any of various complex, spore-bearing fungal structures

fumigant (v. fumigate)—vapor-active chemical used in the gaseous phase to kill or inhibit the growth of microorganisms or other pests

fungicide (adj. fungicidal)—chemical or physical agent that kills or inhibits the growth of fungi

fusiform—narrowing toward the ends

gall—outgrowth or swelling of unorganized plant cells produced as a result of attack by bacteria, fungi, or other organisms

gelatinous—resembling gelatin or jelly

gene—unit within an organism controlling heritable characteristics; bean genes are organized on chromosomes

genetic—relating to heredity; referring to heritable characteristics

geniculate—bent like a knee

genome—set or group of chromosomes

genotype—genetic constitution of an individual, in contrast to its appearance, or phenotype

genus (pl. genera)—group of related species

germ plasm—bearer of heredity material, often loosely applied to cultivars and breeding lines

germ tube—initial hyphal strand from a germinating fungal spore

germinate—to begin growth of a seed or spore

giant cells—multinucleate cells formed by disintegration of cell walls (also called syncytia, in nematode infection)

girdle—to circle and cut through; to destroy vascular tissue, as in a canker that encircles the stem

glazing—glassy appearance

gram-negative, gram-positive—pertaining to bacteria that release or retain, respectively, the violet dye in Gram's stain

gravid—pregnant; bearing eggs

guttation—exudation of watery, sticky liquid from stomata or hydathodes of leaves

habit—form of plant

haploid—having a single complete set of chromosomes (*n* chromosomes)

haustorium (pl. haustoria)—specialized outgrowth (of a stem, root, or mycelium) that penetrates a host plant and extracts nutrients

herbaceous—nonwoody, as a plant or plant part

heterothallic—consisting of two or more mating types

hilum—scar at the point of attachment on a seed or spore

homothallic—consisting of one mating type

host plant—plant on which another organism, particularly a parasite, feeds

host range—the range of plants on which an organism, particularly a parasite, feeds

hyaline—colorless, transparent

hydathode—epidermal leaf structure specialized for secretion or exudation of water

hydrocarbon—chemical compound composed of hydrogen and carbon

hydrolysis—enzymatic cleavage of a chemical bond with addition of water

hygroscopic—pertaining to taking up and retaining water

hymenium—layer of hyphae in an apothecium in which asci and ascospores occur

hyperplasia—abnormal increase in the number of cells in a tissue or organ, resulting in the formation of galls or tumors

hypersensitive—extremely or excessively sensitive; having a type of resistance resulting from extreme sensitivity to a disease

hypertrophy—abnormal increase in the size of cells in a tissue or organ, resulting in the formation of galls or tumors

hypha (pl. hyphae, adj. hyphal)—tubular filament of a fungus

hypocotyl—portion of the stem below the cotyledons and above the root

icosahedron (adj. icosahedral)—a polyhedron having 20 faces

imbibe (n. imbibition)—to take in, especially water during soaking

immune—showing no symptoms of disease when inoculated with pathogen

immunogen—agent inducing production of antibodies

immunogenic—able to induce production of antibodies

in vitro—in glass, on artificial media, or in an artificial environment imperfect state (*see* anamorph)

inclusion body—structure developed within a plant cell as a result of infection by a virus, often useful in identifying the virus

indeterminate—continuing to grow vegetatively while flowering, as beans with a viny habit

indicator plant—plant that reacts to a pathogen (such as a virus) or to an environmental factor with specific symptoms, used to identify the pathogen or determine the effects of the environmental factor

infection—process in which a pathogen enters, invades, or penetrates and establishes a parasitic relationship with a host plant

infection court—site in or on a host plant where infection can occur

infection cushion—organized mass of hyphae formed on the surface of a plant from which numerous infective hyphae develop, common in *Rhizoctonia solani*

infection focus—initial site of infection, generally with reference to a population of plants

infection peg—narrow hypha of a fungal pathogen that penetrates a plant at an infection court

infectious—capable of spreading disease from plant to plant

infective—referring to an organism or virus able to attack a host and cause infection; referring to a vector carrying or containing a pathogen and able to transfer it to a host plant, causing infection

infest—to attack as a parasite (used especially of insects and nematodes); to contaminate, as with microorganisms; to be present in numbers

infundibuliform—shaped like a cone or a funnel

initial inoculum—*see* primary inoculum

inoculate (n. inoculation)—to place inoculum in an infection court

inoculum—pathogen or pathogen part (e.g., spores or mycelium) that infects plants

inoculum potential—energy or potential of inoculum to cause disease

inoperculate—not possessing an operculum

intercalary—inserted within, e.g., located along a hypha as opposed to being located at the end of a hypha

intercellular—between cells

intercrop—to grow two or more crops simultaneously on the same area of land

internode—the portion of a stem between two adjacent nodes

interspecific—between species

interveinal—between (leaf) veins

intracellular—within the cell

intumescence—production of blisters on leaves, usually under conditions of high moisture and restricted transpiration

iron chelate—a ring-shaped chemical compound that binds iron firmly and therefore enables it to be released slowly into the environment

isolate—pure microbial culture, separated from its natural origin

isolation plate—a laboratory dish in which a microbial pathogen is allowed to grow away from its natural substrate in order that a pure culture of the organism can be obtained

isometric—characterized by having three equal axes at right angles

isozyme—one of two or more chemically different substances with the same enzymic activity

landrace—a locally developed strain of a plant

larva (pl. larvae)—juvenile stage of certain animals (e.g., nematodes or beetles) between the embryo and the adult

latent—present but not manifested or visible, as a symptomless infection

lenticular—lens-shaped

lesion—wound or delimited diseased area

life cycle—succession of events in the growth and development of an organism from initiation to death; in microorganisms, may include germination of propagules, growth of colonies, asexual and sexual reproduction, and production of survival structures

local lesion—small, restricted lesion, often a resistance reaction and often the characteristic reaction of differential cultivars to specific pathogens

lodge—to fall over

macroconidium (pl. macroconidia)—the larger and generally more diagnostic type of conidium produced by a fungus that also has microconidia

macrocyclic—referring to a long-cycled rust producing at least one type of binucleate spore (aeciospore or urediniospore) in addition to the teliospore

mating types—morphologically identical forms of a fungus that are sexually compatible and can interact to produce sexual spores

mechanical injury—injury of a plant part by abrasion, mutilation, or wounding

medium (pl. media)—nutritional substrate on which a culture of a microorganism is grown

meiosis—process of nuclear division in which the number of chromosomes per nucleus is halved, e.g., converting the diploid state to the haploid state

meristem (adj. meristematic)—plant tissue characterized by frequent cell division, producing cells that become differentiated into specialized tissues

mesophyll—parenchyma tissue in the abaxial half of a leaf

messenger RNA—a form of RNA that carries information to direct the synthesis of protein

metabasidium—the part of the basidium in which meiosis occurs

microclimate—weather conditions on a small scale, e.g., at the surface of the plant or within a crop

microconidium (pl. microconidia)—the smaller type of conidium produced by a fungus that also has macroconidia

middle lamella—the layer, consisting largely of pectic substances, between the walls of adjacent plant cells

MLO—*see* phytoplasma

molt—to shed a cuticle or body encasement during a growth phase

monilioid—resembling a string of beads

monocotyledon—plant with one cotyledon, such as grasses, grain crops, or corn

monoculture—cultivation of a single crop species

monogenic—pertaining to one gene

morphology—the study of the form of organisms; the form of an organism

mosaic—disease symptom characterized by nonuniform foliage coloration, with a more or less distinct intermingling of normal and light green or yellowish patches, usually caused by a virus; mottle

mottle—disease symptom characterized by light and dark areas in an irregular pattern on a leaf or pod

mucilaginous—viscous, slimy

mulch—layer of material, such as organic matter or plastic, applied to the surface of the soil for purposes such as retention of water and inhibition of weeds

multinucleate—containing many nuclei

multiseptate—having many septa

multitrichous—having many hairs

mutation—heritable genetic change in a cell or plant

mycelium (pl. mycelia)—mass of hyphae constituting the body (thallus) of a fungus

necrogenous—producing necrosis

necrosis (adj. necrotic)—death of tissue, usually accompanied by black or brown darkening

negative staining—staining procedure in which the background is stained and the object is left unstained, so that the object is visualized by contrast with the background

nematicide—agent, usually a chemical, that kills or inhibits nematodes

nesting—formation of a profuse mat of mycelium on fresh bean pods in storage, transit, or market packs

node (adj. nodal)—enlarged portion of a shoot at which leaves or buds are located

nonpersistent transmission—form of virus transmission in which the virus remains transmissible for a short period (e.g., hours or days) while in association with its vector

nonseptate—without cross-walls

nucleus—organelle within a cell containing chromosomes

obclavate—shaped like an upside-down club

obligate parasite—organism that can grow only as a parasite in association with its host plant and cannot be grown in artificial culture media

obovoid—shaped like an upside-down egg

obpyriform—shaped like an upside-down pear

obtuse—blunt

ocherous—earth-colored

oogonium (pl. oogonia)—female egg cell of oomycete fungi

oospore—thick-walled, sexually derived resting spore of oomycete fungi

operculum—lidlike structure at the apex of certain asci that opens when ascospores are ejected from the ascus

ostiole—opening in a fungal fruiting body (e.g., pycnidium or perithecium) through which spores are released

ovary—ovule-bearing portion of a pistil

overwinter—to survive or persist over the winter period

oviposition—egg-laying

ovoid—egg-shaped

ovule—egg of a seed plant, contained within an ovary

paraphysis (pl. paraphyses)—sterile, hairlike hypha in the hymenium

parasite (adj. parasitic)—organism that lives in intimate association with another organism on which it depends for its nutrition

parenchyma—plant tissue composed of rounded or elongate, thin-walled cells

pathogen (adj. pathogenic)—a disease-producing organism

pathogenesis—production of disease

pathotype—general term for a distinctive pathogenic form within a species, less specific than forma specialis or race

pedicel—stalk of a flower

peduncle—stalk of a fruit, e.g., pod

penetrance—the proportion of individuals of a given genotype that express its phenotypic effect in a particular environment

penicillate—shaped like a brush

perennial—plant that survives for several to many years

perfect flower—flower possessing both stamens and pistil

perfect state—*see* teleomorph

persistent transmission—form of virus transmission in which the virus remains transmissible for a prolonged period (e.g., weeks) while in association with its vector

petiole—the stalk of a leaf

pH—negative logarithm of the effective hydrogen ion concentration, a measure of acidity (pH 7 is neutral; values less than pH 7, acidic; values greater than pH 7, alkaline)

phage—a virus parasitic on bacteria

phenotype—the physical properties of an organism produced by the interaction of its genotype with the environment

phialide (adj. phialidic)—end cell of a conidiophore; conidiophore of fixed length with one or more open ends, through which a basipetal succession of conidia develops

phialoconidium (pl. phialoconidia)—conidium produced from a phialide

phloem—food-conducting, food-storing tissue in the vascular system of roots, stems, and leaves

photochemical—pertaining to a chemical reaction influenced by light

photochemical oxidant—any of various highly reactive compounds formed by the action of sunlight on less toxic precursors

photosynthate—product of photosynthesis

photosynthesis—formation of sugars in plants from carbon dioxide and water under the influence of light and facilitated by chlorophyll

phyllody—the state or condition of becoming leaflike

phytoalexin—substance that inhibits the growth of certain microorganisms and that is produced in higher plants in response to a number of chemical, physical, and biological stimuli

phytoplasma—member of a type of prokaryotic, obligate parasites in the class Mollicutes, lacking a cell wall, pleomorphic, and not yet culturable on artificial media

phytotoxic—harmful to plants (usually used to describe chemicals)

phytotoxicity—injury or damage to a plant due to a chemical treatment

phytotoxin—a toxin affecting plants

pinnate—resembling a feather

pistil—the ovule-bearing organ of the plant consisting of the ovary and its appendages (e.g., style or stigma)

pith—loose, spongy tissue in the center of a stem

plano-convexoid—flat and lens-shaped

pleomorphic—variable in shape

plumule—rudimentary shoot of the plant embryo

plurivorous—having many hosts

pollen—male sex cells produced by the anthers of flowering plants

pollination—deposition of pollen on a stigma

polygenic—pertaining to, or governed by, many genes

positive-sense RNA—RNA that can serve directly as messenger RNA

primary infection—the first infection of a plant, usually in the spring by overwintering inoculum

primary inoculum—inoculum, usually from an overwintering source, that initiates disease in the field, as opposed to inoculum that spreads disease during the season

primary leaf—the first true leaf of the bean

plant prokaryote—organism without a distinct nucleus, such as bacteria and phytoplasma

propagule—any part of an organism capable of independent growth

protectant—agent, usually a chemical, applied to a plant surface in advance of a pathogen to prevent infection

proteolysis—breakdown of protein, especially by enzymes

protoplasm—living contents of a cell

proximal—nearest to the point of attachment

pulvinus—a mass of large, thin-walled cells at the base of a petiole that functions in turgor movements of leaves or leaflets

punctate—marked with minute spots

pustule—blisterlike, small erumpent spot, spore mass, or sorus

pycnidiospore—spore (conidium) produced in a pycnidium

pycnidium (pl. pycnidia)—asexual, globose or flask-shaped fruiting body of fungi producing conidia

pycniospore—spore produced within a pycnium

pycnium (pl. pycnia, adj. pycnial)—globose or flask-shaped haploid fruiting body of rust fungi

pyriform—pear-shaped

race—distinct kind of plant pathogen within a species differentiated from other races of the species by the cultivars or varieties of its host plant that it attacks (*see* forma specialis)

raceme—an inflorescence in which the elongated axis bears flowers on short stems in succession toward the apex

radicle—rudimentary root of the plant embryo

receptacle—the structure that bears the reproductive organs, e.g., the stipe and cup of an apothecium

recessive—pertaining to a form of a gene that is not expressed phenotypically when accompanied by a second, dominant, form of the same gene; a recessive character is a phenotypic characteristic of the plant controlled by a recessive gene (*see* dominant)

reniform—kidney-shaped

resistance (adj. resistant)—property of hosts that prevents or impedes disease development

resting spore—temporarily dormant spore, usually thick-walled, capable of surviving adverse environments

reticulate—netlike

Rhizobium—genus of bacteria that live symbiotically with roots of leguminous plants; the symbiosis fixes (converts) atmospheric nitrogen gas into protein

rhizomorph—fungus mycelium arranged in strands, rootlike in appearance

ribosome—RNA-rich organelle in a cell that functions as a site of protein synthesis

rind—hard, outer layer, e.g., of a sclerotium

ring spot—disease symptom characterized by yellowish or necrotic rings enclosing green tissue, as in some plant diseases caused by viruses

RNA—ribonucleic acid

root rot potential—the likelihood that root rot of a particular crop will occur in a field, as determined by testing the soil in the greenhouse or laboratory

rot—softening, discoloration, and often disintegration of succulent plant tissue as a result of fungal or bacterial infection

rotation—growth of different kinds of crops in succession in the same field

rugose (n. rugosity)—wrinkled

russet—scorched or burnt appearance of plant surfaces, especially leaves or pods

sanitation—destruction of infected and infested plants or plant parts

saprophyte—nonpathogenic organism that obtains nourishment from the products of organic breakdown and decay

satellite virus—a virus that accompanies another virus and depends on it for its multiplication

scab—crustlike disease lesion

sclerotium (pl. sclerotia)—hard, usually darkened and rounded mass of dormant hyphae with differentiated rind and medulla and thick, hard cell walls, which permit survival in adverse environments

secondary infection—infection resulting from the spread of infectious material produced after a primary infection or from other secondary infections without an intervening inactive period

secondary inoculum—inoculum produced by infections that took place during the same growing season

secondary organism—organism that multiplies in already diseased tissue but is not the primary pathogen

seed treatment—application of a biological agent, chemical substance, or physical treatment to seed, usually to protect the seed or plant from pathogens or to stimulate germination of the seed or growth of the plant

segregant—member of a population of hybrid progeny that differ phenotypically among themselves

selective medium—a culture medium suitable for the isolation of one to a few kinds of microorganisms

senesce (n. senescence, adj. senescent)—to decline with maturity or age, often hastened by stress from environment or disease

septum (pl. septa, adj. septate)—cross-wall

serology—a method using the specificity of the antigen–antibody reaction for the detection and identification of antigenic substances and the organisms that carry them

seta (pl. setae)—hairlike fungal cell

short-day plant—plant that requires long nights in order to produce flowers

shot-hole—symptom in which small lesions fall out of leaves, giving the leaf the appearance of being hit by buckshot

sieve element—a tube-shaped living cell in the phloem thought to function in the transport of dissolved organic substances in the plant

sign—indication of disease from direct visibility of a pathogen or its parts

sinuous—of a serpentine or wavy form

soil drench—application of a solution or suspension of a chemical to the soil, especially pesticides to manage soilborne pathogens

soil solarization—disease management practice in which soil is covered with polyethylene sheeting and exposed to sunlight, thereby heating the soil and reducing soilborne plant pathogens

sorus (pl. sori)—compact fruiting structure of rust fungi

sp. (pl. spp.)—species (sp. after a genus name refers to an undetermined species; spp. after a genus name refers to several species without naming them individually)

spermagonium (pl. spermagonia)—flask-shaped fungal structure producing sporelike bodies that may function as male gametes (spermatia); pycnium of a rust fungus

spermatium (pl. spermatia)—a sex cell; a nonmotile gamete

sporangiophore—sporangium-bearing body of a fungus

sporangium (pl. sporangia)—fungal structure producing asexual spores, usually zoospores

spore—reproductive body of fungi and some other organisms, containing one or more cells; a bacterial cell modified to survive an adverse environment

sporocarp—spore-bearing fruiting body

sporodochium (pl. sporodochia)—closely grouped cluster of conidiophores

sporulate—to produce spores

stamen—male structure of a flower, composed of a pollen-bearing anther and a filament, or stalk

sterigma (pl. sterigmata)—small, usually pointed protuberance or projection

sterile mycelium—mycelium not producing reproductive structures

stigma—structure on which pollen grains germinate in a pistil

stipe—a stalk

stomate (pl. stomata, adj. stomatal)—structure composed of two guard cells and the opening between them, in the epidermis of a leaf or stem, functioning in gas exchange

strain—a distinct form of an organism or virus within a species, differing from other forms of the species biologically, physically, or chemically

stroma (pl. stromata)—compact mass of mycelium that supports fruiting bodies

stylet—slender, tubular mouthparts in plant-parasitic nematodes or aphids

subgenomic RNA—RNA that is part of the genomic RNA of viruses and acts as messenger RNA for the synthesis of proteins

subpyriform—approximately pear-shaped

subsoiling—deep plowing to break up compacted subsoil

substrate—the substance on which an organism lives or from which it obtains nutrients; chemical substance acted upon, often by an enzyme

susceptible—prone to develop disease when infected by a pathogen

symbiosis (adj. symbiotic)—an intimate relationship of two or more interdependent organisms

symptom—indication of disease by reaction of the host

syn.—synonym

synergism (adj. synergistic)—greater than additive effect of interacting factors

synnema (pl. synnemata)—group of closely united and sometimes fused conidiophores, which bear conidia (syn. coremium) (*see* sporodochium)

systemic—pertaining to a disease in which the pathogen (or a single infection) spreads generally throughout the plant; pertaining to chemicals that spread internally through the plant

taproot—the main root of a bean plant

teleomorph—the sexual form (also called the perfect state or sexual stage) in the life cycle of a fungus, in which sexual spores (ascospores or basidiospores) are formed after nuclear fission

teliospore—thick-walled resting spores produced by some fungi, notably rusts and smuts, that germinate to form a basidium

telium (pl. telia)—sorus that produces teliospores

testa—seed coat

thermal inactivation—elimination of infectivity of a virus by exposure to heat

tissue—group of cells, usually of similar structure, that performs the same or related functions

titer—concentration or infectivity of a virus solution

tolerance—capacity of a plant or crop to sustain disease or endure adverse environment without serious damage or injury

tomentose—covered with a dense mat of hairs

toxin—poisonous substance of biological origin

translocation—movement of water, nutrients, chemicals, or elaborated food materials within a plant

transmit (n. transmission)—to spread or transfer, as in spreading an infectious pathogen from plant to plant or from one plant generation to another

transovarial—via the ovaries

transpiration—water loss by evaporation from leaf surfaces and through stomata

trifoliolate—with three lobes

triturate—to prepare as a powder by rubbing or grinding

truncate—shortened as if being cut off suddenly

ultrastructural—relating to the structure of a cell as seen through an electron microscope

umber—dark earth-colored

unifoliolate—one-lobed, as the primary leaf of common bean

uniseriate—in one row

unitunicate—possessing one layer or wall

urediniospore—repeating vegetative spore of rust fungi

uredinium (pl. uredinia)—fruiting body (sorus) of rust fungi that produces urediniospores

vacuole—generally spherical organelle within a cell bound by a membrane and containing dissolved materials as metabolic precursors, storage materials, or waste products

variegation—pattern of two or more colors in a plant part, as in a green and white leaf

variety—group of closely related plants of common origin and similar characteristics within a species (*see* cultivar)

vascular—pertaining to conductive tissues (xylem and phloem)

vascular bundle—strand of conductive tissue, usually composed of xylem and phloem (in leaves, small bundles are called veins)

vector—agent that transmits inoculum and is capable of spreading disease

vegetative—referring to somatic or asexual parts of a plant, which are not involved in sexual reproduction

vein—small vascular bundle in a leaf

veinbanding—discoloration or chlorosis occurring in bands along leaf veins, setting them off from interveinal tissue, a symptom of virus diseases

veinlet—small branch of a vein ending in the mesophyll of a leaf

vermiform—worm-shaped

vessel—water-conducting cell of xylem

viable (n. viability)—able to germinate, as seeds, fungus spores, sclerotia, etc.; capable of growth

virescence—state or condition of becoming green

virion—mature virus

viroid—the smallest known infectious agent, consisting of nucleic acid and lacking the usual coat of viruses

virulence (adj. virulent)—capacity to cause disease

viruliferous—virus-carrying (usually applied to insects or nematodes)

viviparous—germinating while still attached to the parent plant, e.g., of seed while still in the attached pod

volunteer beans—bean plants developing from seed left in the field from a previous bean crop

water potential—measure of energy available in water

water-soaked—describing plants or lesions that appear wet and dark and are usually sunken and translucent

wilt—loss of freshness or drooping of plants due to inadequate water supply or excessive transpiration; a vascular disease interfering with water utilization

witches'-broom—disease symptom characterized by an abnormal, massed, brushlike development of many weak shoots arising at or close to the same point

xylem—water-conducting, food-storing, supporting tissue of a plant

zonate—marked with zones; having concentric rings, like a target

zoospore—fungus spore with flagella, capable of locomotion in water

Selected References

Agrios, G. N. 2005. Plant Pathology, 5th ed. Academic Press, New York.

Holliday, P. 1998. A Dictionary of Plant Pathology, 2nd ed. Cambridge University Press, Cambridge.

Ulloa, M., and Hanlin, R. T. 2000. Illustrated Dictionary of Mycology. American Phytopathological Society, St. Paul, MN.

Index

Compendium of Bean Diseases

SECOND EDITION

Edited by

Howard F. Schwartz
Colorado State University
Fort Collins

James R. Steadman
University of Nebraska
Lincoln

Robert Hall
University of Guelph
Guelph, Ontario

Robert L. Forster (retired)
University of Idaho R & E Center
Kimberly

APS
PRESS

The American Phytopathology Society

Cover photographs by Howard Schwartz

Reference in this publication to a trademark, proprietary product,
or company name by personnel of the U.S. Department of Agriculture
or anyone else is intended for explicit description only and does not
imply approval or recommendation to the exclusion of others that
may be suitable.

Library of Congress Control Number: 2005930061
International Standard Book Number: 0-89054-327-5

Printed in China on acid-free paper.

The American Phytopathological Society
3340 Pilot Knob Road
St. Paul, Minnesota 55121, U.S.A.